my **revisi⏻n** notes

Edexcel AS/A-level

GEOGRAPHY

Michael Witherick
Dan Cowling

HODDER
EDUCATION
AN HACHETTE UK COMPANY

Photo credit:
p.147 © Romeo Gacad/AFP/Getty Images

Hachette UK's policy is to use papers that are natural, renewable and recyclable products and made from wood grown in sustainable forests. The logging and manufacturing processes are expected to conform to the environmental regulations of the country of origin.

Orders: please contact Bookpoint Ltd, 130 Park Drive, Milton Park, Abingdon, Oxon OX14 4SE. Telephone: (44) 01235 827720. Fax: (44) 01235 400454. Email education@bookpoint.co.uk Lines are open from 9 a.m. to 5 p.m., Monday to Saturday, with a 24-hour message answering service. You can also order through our website: www.hoddereducation.co.uk

ISBN: 978 1 4718 8674 4

First published in 2017 by
Hodder Education,
An Hachette UK Company
Carmelite House
50 Victoria Embankment
London EC4Y 0DZ
www.hoddereducation.co.uk

Impression number   10 9 8 7 6 5 4 3 2 1

Year   2021 2020 2019 2018 2017

Cover photo reproduced by permission of Tim UR/Fotolia
Typeset in Bembo Std Regular 11/13 by Integra Software Services Pvt. Ltd., Pondicherry, India
Printed in Spain

A catalogue record for this title is available from the British Library.

Get the most from this book

Everyone has to decide his or her own revision strategy, but it is essential to review your work, learn it and test your understanding. These Revision Notes will help you to do that in a planned way, topic by topic. Use this book as the cornerstone of your revision and don't hesitate to write in it — personalise your notes and check your progress by ticking off each section as you revise.

Tick to track your progress

Use the revision planner on pages 4 and 5 to plan your revision, topic by topic. Tick each box when you have:
● revised and understood a topic
● tested yourself
● practised the exam questions and gone online to check your answers and complete the quick quizzes.

You can also keep track of your revision by ticking off each topic heading in the book. You may find it helpful to add your own notes as you work through each topic.

Features to help you succeed

Key concept

Essential concepts are explained more fully to aid understanding.

Exam tips and summaries

Expert tips are given throughout the book to help you polish your exam technique in order to maximise your chances in the exam. The summaries provide a quick-check bullet list for each topic.

Typical mistakes

The author identifies the typical mistakes students make and explain how you can avoid them.

Now test yourself

These short, knowledge-based questions provide the first step in testing your learning. Answers are at the back of the book.

Definitions key words

Clear, concise definitions of essential key terms are provided where they first appear. Key words

from the specification are highlighted in colour throughout the book.

Revision activities

These activities will help you to understand each topic in an interactive way.

Exam practice

Practice exam questions are provided for each topic. Use them to consolidate your revision and practise your exam skills.

Synoptic theme

Over-arching themes help you to make links between different geographical ideas and concepts.

Online

Go online to check your answers to the exam questions and try out the extra quick quizzes at **www.hoddereducation.co.uk/myrevisionnotes**

My revision planner

REVISED   TESTED   EXAM READY

Exam practice answers and quick quizzes at www.hoddereducation.co.uk/myrevisionnotes

Countdown to my exams

6–8 weeks to go

- Start by looking at the specification — make sure you know exactly what material you need to revise and the style of the examination. Use the revision planner on pages 4 and 5 to familiarise yourself with the topics.
- Organise your notes, making sure you have covered everything on the specification. The revision planner will help you to group your notes into topics.
- Work out a realistic revision plan that will allow you time for relaxation. Set aside days and times for all the subjects that you need to study, and stick to your timetable.
- Set yourself sensible targets. Break your revision down into focused sessions of around 40 minutes, divided by breaks. These Revision Notes organise the basic facts into short, memorable sections to make revising easier.

REVISED ☐

2–6 weeks to go

- Read through the relevant sections of this book and refer to the exam tips, summaries, typical mistakes and key terms. Tick off the topics as you feel confident about them. Highlight those topics you find difficult and look at them again in detail.
- Test your understanding of each topic by working through the 'Now test yourself' questions in the book. Look up the answers at the back of the book.
- Make a note of any problem areas as you revise, and ask your teacher to go over these in class.
- Look at past papers. They are one of the best ways to revise and practise your exam skills. Write or prepare planned answers to the exam practice questions provided in this book. Check your answers online and try out the extra quick quizzes at **www.hoddereducation.co.uk/myrevisionnotes**
- Use the revision activities to try out different revision methods. For example, you can make notes using mind maps, spider diagrams or flash cards.
- Track your progress using the revision planner and give yourself a reward when you have achieved your target.

REVISED ☐

One week to go

- Try to fit in at least one more timed practice of an entire past paper and seek feedback from your teacher, comparing your work closely with the mark scheme.
- Check the revision planner to make sure you haven't missed out any topics. Brush up on any areas of difficulty by talking them over with a friend or getting help from your teacher.
- Attend any revision classes put on by your teacher. Remember, he or she is an expert at preparing people for examinations.

REVISED ☐

The day before the examination

- Flick through these Revision Notes for useful reminders, for example the exam tips, topic summaries, typical mistakes and key terms.
- Check the time and place of your examination.
- Make sure you have everything you need — extra pens and pencils, tissues, a watch, bottled water, sweets.
- Allow some time to relax and have an early night to ensure you are fresh and alert for the examinations.

REVISED ☐

My exams

A-level Geography Paper 1

Date:..

Time:..

Location:..

A-level Geography Paper 2

Date:..

Time:..

Location:..

A-level Geography Paper 3

Date:..

Time:..

Location:..

1 Tectonic processes and hazards

Tectonic hazards are earthquakes and volcanic eruptions. Also included under the heading are secondary hazards, such as tsunamis.

Spatial variations in the tectonic hazard risk

The global distribution of tectonic hazards

Key concepts

Plate boundaries are recognised by plate tectonic theory as being of three types:

- Convergent (destructive): occur where two tectonic plates are moving together. Where a dense oceanic plate collides with a less dense continental plate, the former is thrust underneath the latter, forming a subduction zone. Mountain building and volcanic eruptions are the outcomes.
- Divergent (constructive): the moving apart of the plates creates rifts filled with new crustal material from volcanic eruptions.
- Conservative (transform): here two crustal plates slide past each other. The friction often triggers earthquakes.
- Collision: two continental plates collide and crush against each other, pushing up mountains.

Earthquakes

The global distribution of tectonic hazards is far from random. The main earthquake zones occur along **plate boundaries**, particularly convergent and conservative ones (Figure 1.1). Occasionally earthquakes occur in the middle of tectonic plates (intra-plate earthquakes).

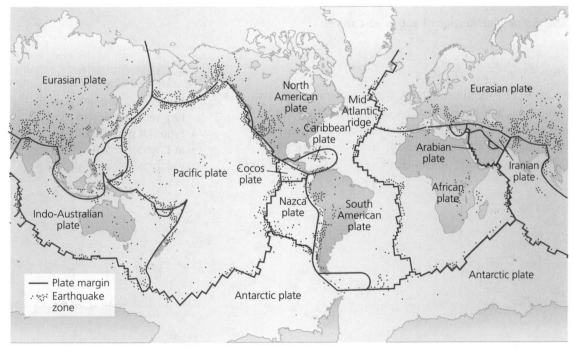

Figure 1.1 The global distribution of earthquakes

Volcanoes

There are about 500 active volcanoes around the world. Figure 1.2 shows that a significant number of these are located in the 'Ring of Fire' around the Pacific Ocean. Most volcanoes occur near plate boundaries, but there are also **hotspot volcanoes**.

> **Hotspot volcanoes:** These are found in the middle of tectonic plates and are thought to be fed from the underlying mantle (a thick layer of high-density rocks lying between the Earth's crust and its molten core). These volcanoes occur where the mantle is unusually thin and hot. The summits of the Hawaiian islands are classic examples.

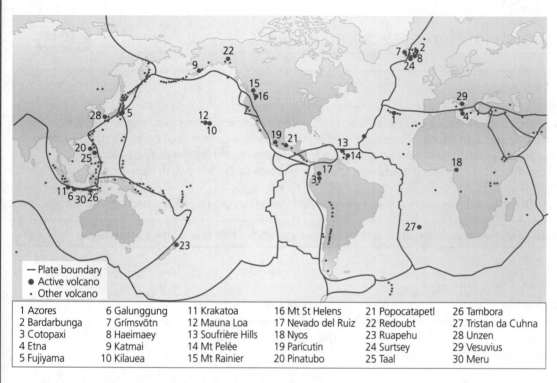

1 Azores	6 Galunggung	11 Krakatoa	16 Mt St Helens	21 Popocatapetl	26 Tambora
2 Bardarbunga	7 Grímsvötn	12 Mauna Loa	17 Nevado del Ruiz	22 Redoubt	27 Tristan da Cuhna
3 Cotopaxi	8 Haeimaey	13 Soufrière Hills	18 Nyos	23 Ruapehu	28 Unzen
4 Etna	9 Katmai	14 Mt Pelée	19 Parícutin	24 Surtsey	29 Vesuvius
5 Fujiyama	10 Kilauea	15 Mt Rainier	20 Pinatubo	25 Taal	30 Meru

Figure 1.2 **The global distribution of active volcanoes**

Tsunamis

Tsunamis are caused by submarine shock waves generated by earthquakes or volcanic eruptions, and have a wide global distribution (Figure 1.3). They are most commonly experienced around the coastlines of the Pacific Ocean. Tsunamis are are potentially most devastating where a gently sloping continental shelf allows them to build to great heights.

Exam practice answers and quick quizzes at **www.hoddereducation.co.uk/myrevisionnotes**

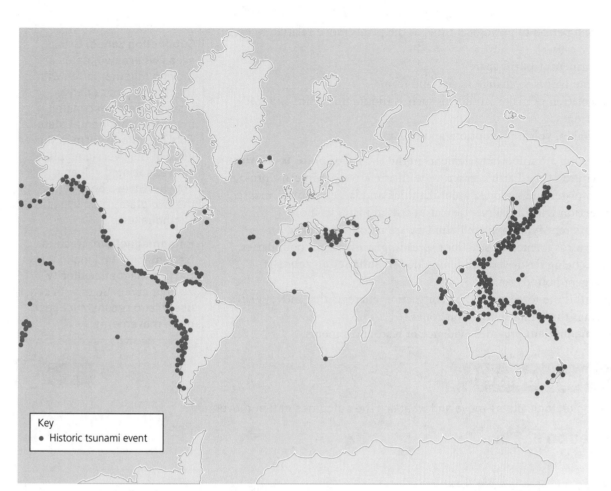

Figure 1.3 **Notable tsunami events since 1900**

Now test yourself

1 Name two locations outside the Pacific Rim where there have been a number of notable tsunamis.

Answer on p. 215

Theoretical frameworks

The theory

> **Key concept**
>
> **Plate tectonics** theory views the Earth's crust as consisting of a number of mobile yet rigid elements (plates). These plates are of two different types:
> - thin crust underlying the ocean basins
> - thicker crust underlying the continents.
>
> The low density of the thick continental crust allows it to 'float' on the much higher-density mantle below. Heat derived from the Earth's molten core rises within the mantle to create convection currents which, in turn, move the tectonic plates.

Over a long period of geological time, as the plates move relative to each other, they cause:

- the continents to drift apart
- the ocean basins to change in size and form
- the formation of major landforms such as mountain chains and mid-ocean ridges
- earthquakes, volcanic eruptions and tsunamis.

The following are important elements in the theory of **plate tectonics**:

- The nature of the Earth's structure with a relatively thin crust broken up into plates and wrapped around a thick and largely molten mantle.
- Convection within the mantle causes crustal plates to move.
- Four different types of plate boundary are recognised (page 7).
- New crust is formed by sea-floor spreading at divergent boundaries.
- Crust is being destroyed and remoulded in **subduction zones** at convergent boundaries.
- Slab pull is the force created by convection currents that moves plates and drags them into subduction zones.
- **Paleomagnetism** provides evidence of plate movements.

> **Subduction zones:** Broad areas where two plates are moving together, often with the thinner, more dense oceanic plate descending beneath a continental plate. Fold mountains form at the edge of the overriding plate, with associated volcanic activity. Stress between the two plates also triggers earthquakes.
>
> **Paleomagnetism:** Results from magma locking in the Earth's magnetic polarity when it cools. Scientists can use this to reconstruct past plate movements.

Now test yourself

TESTED

2 How do tectonic plates move and what are the outcomes of that movement?

Answer on p. 215

The type and magnitude of event

The type of tectonic event is largely determined by the type of plate boundary. The convergent boundary is the most productive of both earthquakes and volcanic eruptions, followed by the divergent boundary. The conservative boundary produces only earthquakes.

Science has still to discover what determines the magnitude of a tectonic event. The Benioff zone is thought to be important in the case of some earthquakes. This is the boundary between an oceanic plate that is undergoing subduction and an overriding continental plate. It is a sloping plane and stresses are built up as the cold oceanic plate sinks into the hot mantle. The zone produces earthquakes, but why are some of those earthquakes more powerful than others?

Now test yourself

TESTED

3 Is event magnitude more important than event location? Give your reasons.

Answer on p. 215

Physical processes behind tectonic hazards

REVISED

Earthquakes

Earthquakes are caused by sudden movements of the Earth's crust relatively close to the surface, usually along a pre-existing fault. The movement is the outcome of a gradual build-up of tectonic pressure and then its sudden release. The sudden movement creates vibrations (seismic waves) of three different kinds:

- P (fast)
- S (slower)
- L (surface).

The **hypocentre** of an earthquake, sometimes referred to as the focus, is the point of origin within the Earth's crust where the pressure is released – the point of rupture. The **epicentre** is the point on the Earth's surface directly above the hypocentre. It is the surface location where the shock waves are likely to be strongest.

The overall severity of an earthquake is determined by the amplitude and frequency of these waves. The S and L waves are more destructive than the P waves. They cause crustal fracturing, ground shaking and three secondary hazards:

- Liquefaction: this affects loose rock and sediment. The seismic waves trigger the ground to lose its load-bearing capacity, causing large buildings to settle into the ground, tilt and possibly collapse.
- Landslides: these occur where slopes are weakened by seismic waves and slide under the influence of gravity.
- Tsunamis (see below).

Now test yourself

TESTED

4 What are seismic waves? What is the difference between the hypocentre and the epicentre of an earthquake?

Answer on p. 215

Tsunamis

These waves are potentially the most lethal of the secondary earthquake hazards. Out at sea they do not represent a hazard since they are low in height and generally go unnoticed. It is only as they approach a coastline and the sea becomes shallower that they grow in height.

The impact of a tsunami depends on a number of physical and human factors:

- duration of the event
- wave amplitude and distance travelled
- depth and gradient of the offshore zone
- degree of coastal protection provided by mangroves and coral reefs
- timing of the event – night or day
- quality of early-warning systems
- density of population and degree of development close to the coastline.

Volcanoes

The primary hazards of a volcano are:

- Pyroclastic flows: the frothing of magma at the vent produces bubbles that burst explosively to eject hot and poisonous gases as well as hot, fine materials. Clouds formed of these gases and materials are most lethal when they roll down the sides of a volcano.
- Tephra (ash falls): these are rock fragments ejected into the atmosphere and ranging in size from 'bombs' to fine dust. The accumulation of tephra on roofs starts fires and causes buildings to collapse.

- Lava flows: these are flows of molten rock, often fast moving and lethal.
- Volcanic gases: these are mixed gases emitted during explosive eruptions. The carbon dioxide is particularly dangerous.

The secondary hazardous impacts of volcanoes are:

- lahars: mudflows created by the combination of heavy rain on slopes covered by fine volcanic material
- jokulhaups: catastrophic floods caused by volcanic eruptions beneath glaciers.

Exam tip

Remember that compared with other hazards, such as earthquakes and tsunamis, volcanoes have historically killed far fewer people. An important factor is that there is often some form of advanced warning of an eruption.

Revision activity

Make sure that you have a located, recent example of each of the three tectonic hazards, together with the date of the event and some indications of the scale of the human impact.

Now test yourself

TESTED

5 Which of the primary hazards of a volcanic eruption is potentially the most lethal? Give your reasons.

Answer on p. 215

Tectonic hazards become disasters

Vulnerability, risk, resilience and disaster

REVISED

Vulnerability and risk are key factors in turning **hazards** into **disasters**.

Key concepts

Vulnerability relates to the ability of a community to cope with the impacts of a hazard. That ability is determined by a range of factors, from the quality of warning systems and emergency responses to the level of development and settlement density. So it is argued that a developed country, with good governance and access to technology and relevant resources, is less vulnerable to the same hazard than a developing country. The likelihood of that hazard becoming a disaster is reduced.

Risk is the exposure of people to a hazardous event. It relates to the probability of a hazard leading to a loss of life and/or livelihoods. The assessment of risk is complicated by many factors, including:

- the perceptions of individuals and communities
- the unpredictability of hazards, with people being caught out by the timing or magnitude of a tectonic event
- the lack of alternatives – people continue to live in hazardous areas because they have no other options
- the fact that the benefits of hazardous location may outweigh risks involved in staying there
- acceptance of the risk that something might happen.

Hazards: These are natural events that threaten or actually cause injury and death, as well as damage and destruction to property.

Disasters: These occur when hazards have a significant impact on vulnerable populations. Officially, a hazard becomes a disaster when 100 or more people are killed and/or 100 or more people are affected.

Typical mistake

Tectonic events by themselves are not hazards. They become hazardous only when they adversely impact on people, their settlements and livelihoods.

Typical mistake

The terms hazard and disaster are often taken to mean the same thing. In fact, they mean very different things.

Exam practice answers and quick quizzes at **www.hoddereducation.co.uk/myrevisionnotes**

The hazard–risk formula involves the components that influence the amount of risk a community is taking with a particular type of hazard:

$$\text{risk} = \frac{\text{hazard} \times \text{exposure} \times \text{vulnerability}}{\text{manageability}}$$

The pressure and release (PAR) model adopts a slightly different approach to the assessment of the risk of a hazard becoming a disaster. It sees disaster as occurring at the intersection of two processes:
● those generating vulnerability
● those of the natural hazard event (Figure 1.4).

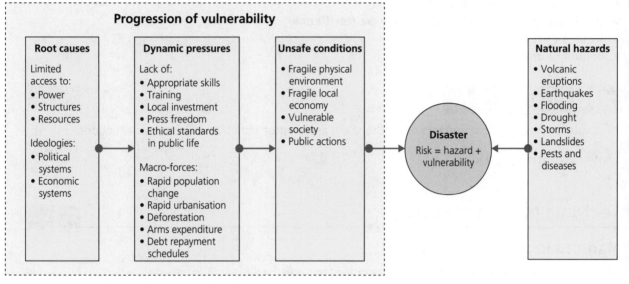

Figure 1.4 The pressure and release model

Root causes create vulnerability through different pressures such as inadequate training and poor government. Dynamic pressures produce unsafe conditions (environmentally and socially) for the most vulnerable people.

One other component that needs to be taken into account when weighing up the risk of a hazard becoming a disaster is resilience.

Key concept

Resilience is the ability of a community or country exposed to hazards to resist, absorb and recover from the impacts of a hazard. Resilience can often help prevent a hazard from becoming a disaster.

Now test yourself

TESTED

6 What is the difference between a hazard and a disaster?
7 What is the difference between vulnerability and resilience?

Answers on p. 215

Impacts

The economic and social impacts of tectonic hazards vary considerably:
● over time
● from place to place
● from minor nuisances to major disasters.

The impacts of earthquakes and their secondary hazards are generally much greater than those of volcanic eruptions. The concentration of active volcanoes in relatively narrow belts means that only a small land area lies in close proximity. It is estimated that less than 1 per cent of the world's population is likely to suffer the impacts of a volcanic eruption, whereas the estimate for earthquakes is 5 per cent.

The economic impacts of a tectonic hazard are roughly proportional to the land area exposed to the particular hazard. But there are other factors involved, such as:

● level of development and per capita GDP
● total number of people affected
● speed of recovery from the hazardous event (**resilience**)
● degree of urbanisation
● amount of uninsured losses.

Now test yourself

TESTED

8 Explain why the global impacts of earthquakes are greater than those of volcanic eruptions.

Answer on p. 215

Tectonic hazard profiles

REVISED

Magnitude and intensity

Magnitude and **intensity** are important aspects of tectonic hazards. Observations and measurements are converted to mathematical scales. Of four widely used scales, three relate to earthquakes and one to volcanic activity (Table 1.1).

Table 1.1 Scales used for two different types of tectonic hazard

	Hazard	Scale	Overview
Richter scale	Earthquake	0–9	A measurement of the height (amplitude) of the waves produced by an earthquake. The Richter scale is an absolute scale – wherever an earthquake is recorded, it will measure the same on the Richter scale.
Mercalli scale (modified)	Earthquake	I–XII	Measures the experienced impacts of an earthquake. It is a relative scale because people experience different amounts of shaking in different places. It is based on a series of key responses, such as people awakening, the movement of furniture and damage to structures.
Moment magnitude scale (MMS)	Earthquake	0–9	A modern measure used by seismologists to describe earthquakes in terms of energy released. The magnitude is based on the 'seismic moment' of the earthquake, which is calculated from the amount of slip on the fault, the area affected and an Earth-rigidity factor. The US Geological Survey (USGS) uses MMS to estimate magnitudes for all large earthquakes.
Volcanic explosivity index (VEI)	Volcanic eruption	0–8	A relative measure of the explosiveness of a volcanic eruption, which is calculated from the volume of products (ejecta), height of the eruption cloud and qualitative observations. Like the Richter scale and the MMS, the VEI is logarithmic: an increase of one index indicates an eruption that is ten times as powerful.

None of the scales shown in Table 1.1 is perfect. For example, they do not take into account the duration of the hazard, the physical exposure or the vulnerability and resilience of the affected communities.

Hazard profiles

Given a set of criteria, it is possible to compile a **tectonic hazard profile** which can then be compared with the profiles of other events. Figure 1.5 shows one style in which the characteristics of earthquakes at two different plate boundaries are compared.

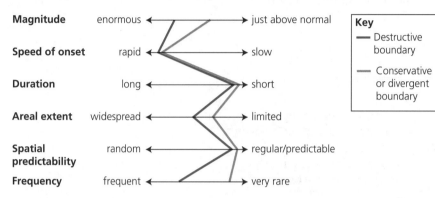

Figure 1.5 Earthquake hazard profiles

The hazard profile is not the only factor that determines the social and economic impacts of an event. For example, it is generally assumed that the impacts of tectonic hazards are likely to be greater in developing countries because of higher levels of vulnerability and lower levels of resilience. The significance of development and other factors is explored in a little more detail in the following section.

> **Revision activity**
>
> Make sure you have notes detailing the impacts of one specific tectonic hazard event in each of three different geographical locations: a developing country, an emerging country and a developed country.

Now test yourself

TESTED

9 What is the value of compiling hazard profiles?

Answer on p. 215

The importance of development and governance

REVISED

The point has already been made that vulnerability and resilience often correlate with development – the former inversely and the latter directly. This is particularly the case with the level of economic development. Economic development gives communities and countries access to the resources, organisations and technology needed to cope with hazard events. With increasing income, people are better able to ensure their own safety by living in 'safe' locations and in 'hazard-proofed' property.

> **Typical mistake**
>
> In the reporting of earthquake magnitude today the numbers given by the media refer to the Richter scale. In fact, it is the MMS that is used.

> **Tectonic hazard profile:** A technique used to try to understand the physical characteristics of different types of hazard, such as earthquakes, volcanic eruptions and tsunamis.

> **Revision activity**
>
> Note the six different criteria that are used in building up the hazard profile. These are critical.

> **Exam tip**
>
> Any assessment of the risks posed by tectonic hazards must identify the nature and magnitude of the hazard, the number of people at risk, the amount of economic investment, the vulnerability of society, and the society's ability to respond to and mitigate the impacts of the hazard.

However, there are non-economic aspects of development that are also significant:

- Access to education: education means that people can be made more aware of the hazard risks of living where they do and of what to do in the event of a hazard.
- Access to healthcare: the better people's health, the better they are at withstanding the health and food risks resulting from the hazard.
- Housing: poorly built housing is usually unable to withstand earthquake shock waves, leading to serious injury and death.
- Governance: the quality of governance can be quite critical (see below).

Governance

> ### Key concept
>
> **Governance** is the way a country, city, community, company, etc. is run by the people in control. It is based on three concepts: authority, decision making and accountability. Good governance embodies the recognition and practice of a range of principles, such as transparency, the rule of law, equity, consensus and participation.

Poor **governance** in the form of corrupt local and national government and weak political organisation increases hazard vulnerability in two ways:

- By failing to invest properly in infrastructure that might mitigate the impacts of a tectonic hazard, for example failing to invest in warning systems, 'hazard-proofing' buildings, etc.
- By being ill-prepared to deal with the emergency situation immediately following a hazard.

> ### Synoptic theme
>
> Clearly, when it comes to governance, national and local governments are the top players – their transparency and efficiency are paramount.

But governance is not only about political authority. There are other people and organisations (**stakeholders**), both public and private, that have a role to play in good governance. They do so by observing the key principles of good governance, such as accountability and participation in responsible decision making.

> **Stakeholders:** Individuals, communities, organisations, businesses and governments with a specific interest in a situation – in this instance, in hazard risk and hazard mitigation.

> ### Now test yourself
>
> TESTED
>
> 10 Why is good governance so important in the context of tectonic hazards?
>
> Answer on p. 215

Geographical factors

Finally, there various geographical factors that can increase hazard vulnerability. These include:

- population density: the higher the density, the more people at risk
- urbanisation: the more people and businesses are concentrated in cities, the higher the risk and vulnerability

Exam practice answers and quick quizzes at **www.hoddereducation.co.uk/myrevisionnotes**

- isolation and inaccessibility: this is particularly critical in the immediate aftermath of a hazard event when there is an urgent need to provide emergency aid
- community spirit: a strong spirit can certainly help boost morale and the collective wish to survive the hazard.

Contrasting locations

The specification requires you to make a comparative study of tectonic hazard events in three different geographical contexts (a developed country, an emerging country and a developing country), focusing on the significance of development. The student book published by Hodder Education contains studies of three earthquake events, in New Zealand (2010–2011), Iran (2003) and Nepal (2015). Despite considerable damage and destruction, only in New Zealand did the earthquake event not become a disaster.

In the case of the Iran (Bam) and Nepal earthquakes, two factors which turned them into disasters were:
- the poorly constructed and vulnerable housing and other buildings
- a poor emergency response – lack of equipment and specialised medical and rescue training. In Bam's case, the situation was not helped by the destruction of the three main hospitals. In Nepal, the problem was made worse by the inaccessibility of the stricken areas – remote mountainous locations with roads rendered impassable by huge landslides.

The management of tectonic hazards and disasters

Trends and patterns

Compared with other natural hazards, few tectonic hazards develop into disasters. Tectonic hazards cannot be prevented. Neither can their spatial occurrence be changed.

Figure 1.6 shows that the annual number of tectonic (geophysical) hazard events involving losses (life, property, etc.) has remained fairly stable compared with meteorological and hydrological hazard events. In other words, such events seem to occur regularly.

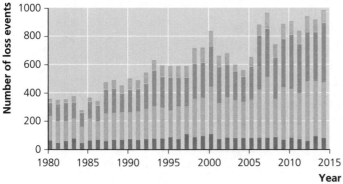

Figure 1.6 The number of hazard loss events, by type (1980–2014)

> **Exam tip**
>
> Remember that the significance of these factors is always conditioned by the magnitude and intensity of the tectonic hazard event.

> **Revision activity**
>
> It is important that you have some notes about the impacts of the same type of hazard in the three different contexts. Did you study those place studies in your student book or did your teacher introduce you to some different examples? Whichever is the case, be sure to brush up on the details.

> **Revision activity**
>
> Make notes on the trends in the four types of hazard event since 1980.

Key
- Climatological events (extreme temperatures, drought, wildfire)
- Hydrological events (flood, mass movement)
- Meteorological events (tropical storm, extra-tropical storm, convective storm, local storm)
- Geophysical events (earthquake, tsunami, volcanic activity)

Table 1.2 Total number of tectonic disasters grouped by country's level of development (2004–2013)

Hazard	Very high HDI	High HDI	Medium HDI	Low HDI	TOTAL
Earthquakes and tsunamis	41	71	121	36	269
Volcanic eruptions	5	12	30	10	57

Table 1.2 confirms that volcanic eruptions cause far fewer disasters than earthquakes. But the smaller numbers also reflect the fact that there are many more earthquakes than volcanic eruptions occurring during the course of a year. The table does not show the expected simple pattern when the number of tectonic disasters is analysed in terms of the human development index (HDI) of the countries in which they occurred. Is the fact that most tectonic disasters occurred in medium HDI countries (emerging countries) really explained by the level of development? Might it not be that their incidence reflects the fact that these countries happen to be located in unstable parts of the Earth's crust?

All hazard data needs to be treated with some caution for a number of reasons:
- There is no universally agreed definition of a disaster.
- Smaller events in remote locations are often under-recorded.
- Disaster deaths and damage are sometimes under-recorded for political reasons.
- **Mega-disasters**, such as the Asian tsunami (2004), the Eyjafjallajökull disaster (2010) and the Japanese tsunami (2011), can distort trends in disaster losses.

> **Mega-disasters:** These result from tectonic hazards and show several diagnostic features:
> - They are large-scale in terms of the area involved and their economic and human impacts.
> - They pose huge challenges, particularly at the emergency stage.
> - They usually require substantial amounts of international disaster aid.

Exam tip

It is recommended that you have to hand some of the details of one mega-disaster.

It needs to be understood that there are parts of the world at risk from more than one type of hazard. So it is appropriate to think in terms of multiple-hazard zones. Figure 1.7 plots zones regularly experiencing three different types of hazard. Locations where the three distributions overlap can be identified as **hazard hotspots**. Table 1.3 shows the eight countries most exposed to multiple hazards. In the case of tectonic hazards, it should be noted that their impacts are often aggravated by hydro-meteorological events which encourage liquefaction and landslides on slopes weakened by earthquakes.

> **Hazard hotspots:** These locations are extremely disaster prone for a number of reasons. Notable is the fact that they experience more than one type of natural hazard.

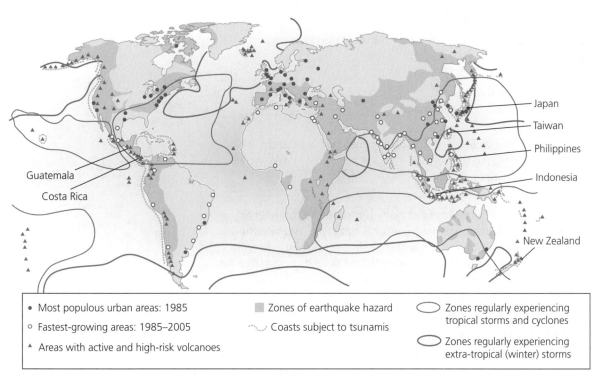

Figure 1.7 The global pattern of multiple hazards

- Most populous urban areas: 1985
- Fastest-growing areas: 1985–2005
- Areas with active and high-risk volcanoes

- Zones of earthquake hazard
- Coasts subject to tsunamis

- Zones regularly experiencing tropical storms and cyclones
- Zones regularly experiencing extra-tropical (winter) storms

Japan
Taiwan
Philippines
Indonesia
New Zealand
Guatemala
Costa Rica

Table 1.3 The countries most exposed to multiple hazards

Country	Total area exposed (%)	Population exposed (%)	Number of different hazards the country is exposed to
Taiwan	73.1	73.1	4
Costa Rica	36.8	41.1	4
Vanuatu	28.8	20.5	3
Philippines	22.3	36.4	5
Guatemala	21.3	40.8	5
Ecuador	13.9	23.9	5
Chile	12.9	54.0	4
Japan	10.5	15.3	4

Revision activity

Identify two locations which Figure 1.7 shows as being exposed to four different hazards.

Prediction and management

REVISED

Prediction

Predicting the occurrence of tectonic hazards is an obvious starting point in any attempt to reduce the deaths and destruction they cause. Research has now made us more aware of tell-tale signs of an imminent volcanic eruption. Earthquakes are altogether more difficult to predict. However, it is beginning to look as if there might be some early warning signs. The key to success is being able to detect those areas of particular stress in the Earth's crust that trigger earthquakes.

Hazard management cycle

The hazard management cycle involves a number of stages once the hazard has struck:

1 Emergency response
2 Initial recovery (rehabilitation)
3 Reconstruction (including **mitigation**)
4 Return to normality
5 Appraisal of the lessons learned during the hazard event and implementation of remedial actions
6 Improving **preparedness**

> **Mitigation:** Any action taken to reduce or eliminate the long-term risk to human life and property from natural hazards. Those actions are largely the outcome of stage 5 in the hazard management cycle (Figure 1.8). However, they are also likely to be taken during stage 3, sometimes referred to as adjustment or adaptation.
>
> **Preparedness:** Educating people about what they should do in an emergency (where to seek shelter, how to assist others) as well as improving warning systems and training, and equipping rescue teams. It is focused on the emergency stage immediately following a hazard.

The next section will provide some examples of the types of action falling under these headings.

The choice of response depends on a complex series of interlinked physical and human factors, such as those shown in Figure 1.8. This figure does not show the critical stages 5 and 6 in the 'disaster-free period'.

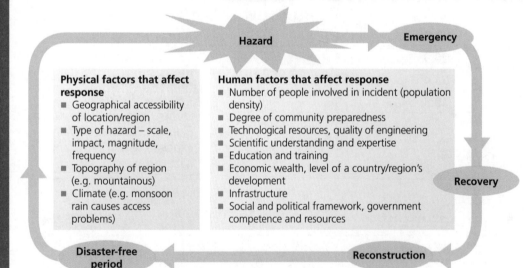

Figure 1.8 The range of factors affecting the response to hazards

Synoptic theme

Key players in the prediction of tectonic disasters are the scientists. But the accuracy of any forecasting depends on the type and location of the tectonic hazard.

Revision activity

Make a list of the factors affecting the response to a hazard.

Park's disaster response curve

Park's disaster response curve is a model that can be used to help analyse the timeline between when a hazard strikes and when a place or community returns to normal life (Figure 1.9). The model recognises five stages which are a near match with the stages in the hazard management cycle. The model allows the response curves of different hazard events to be compared.

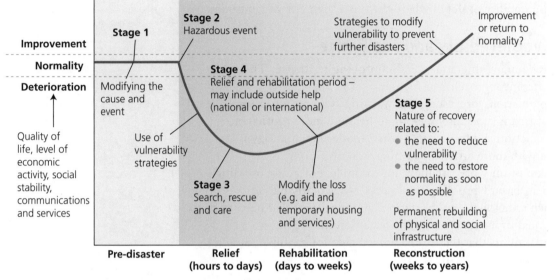

Figure 1.9 Park's model: the disaster response curve

Synoptic theme

Particularly critical in the mitigation of hazard events are the actions involved in planned emergency procedures. All too often the planning of such actions has been found to be wanting, with unnecessary delays and confusion as to who does what.

Exam tip

It is tempting to think that the shorter the timeline of the five stages of the hazard, the greater the resilience of the striken community. It is also tempting to think that the response curves will generally be shorter in developed countries compared with developing countries. However, remember that the nature of the response curve is also going to be greatly affected by the magnitude of the hazard event.

Mitigation and adaptation strategies

REVISED

There are three basic actions that can be undertaken to mitigate the impacts of a tectonic hazard:
- Modify the hazard event.
- Modify both vulnerability and resilience.
- Modify the potential financial losses.

These actions should be informed by risk assessments and hazard predictions.

Exam tip

It is important to be aware of these three different mitigation strategies.

Modifying the hazard event

There is little that can be done under this heading. No technologies are yet capable of preventing tectonic disturbances. The best that can be achieved is to modify, i.e. reduce the hazard impacts by mitigating or adaptive actions under the next heading. However, mention might be made here of:

- strengthening coastal defences against tsunamis
- diverting or chilling lava flows
- increasing the stability of slopes where there is a high risk of landslides.

Modifying vulnerability and resilience

This involves reducing vulnerability and improving resilience. There are a number of different approaches or focuses:

- Improving prediction, forecasting and warning systems: for example, scientific research is constantly seeking to become more proficient in hazard prediction and forecasting, while modern technology is providing us with more efficient warning systems.
- Improving community preparedness: for example, enforcing building codes aimed at 'hazard-proofing' structures, particularly public buildings such as hospitals, police stations and pipelines, that need to be fully operational immediately after the hazard event.
- Changing behaviours that reduce the hazard risk: for example, moving people away from high-risk areas.

Synoptic theme

Key players in hazard mitigation and adaptation are planners (avoiding developments in hazardous locations) and engineers (hazard-proofing buildings and locations; possibly modifying hazard events).

Modifying losses

One obvious way of reducing the financial costs of the losses to the stricken community or country is through insurance. Insurance is expensive but in most instances the actual costs of repair and reconstruction will be significantly more. Of course, the insurance industry has to assess:

- the level of risk in a particular location
- the probability of a hazard of a certain magnitude happening
- the market value of the properties to be insured
- the likely costs of repair or reconstruction.

With earthquakes, seismologists are working with risk analysts to help the insurance industry calculate premiums and risk. Computer simulations are used to estimate the probability of damage from different scales of earthquake event. However, with volcanic eruptions, there is a greater confidence in the assessments of risk and the potential scale of damage.

Now test yourself

TESTED

11 Suggest why insurance for hazard damage is so expensive.

Answer on p. 215

Disaster aid is another way in which hazard losses might be reduced, particularly during the emergency and early recovery stages. The funding for disaster aid has two sources:

- donations by governments to intergovernmental organisations like the UN
- private donations to voluntary organisations and charities, such as the Red Cross, Oxfam and MSF.

Disaster aid is often criticised, largely on the grounds that national and local distribution systems are often inefficient or corrupt, it does not encourage self-help and it does not encourage a more bottom-up management of disasters at a local level.

The Sendai Framework (2015) has set out four priorities in disaster management:

1 Understand the disaster risk.
2 Ensure a strengthening of governance to manage the hazard risk.
3 Invest in improving resilience and disaster preparedness.
4 'Build back better' in the recovery, rehabilitation and reconstruction stages.

It is also recognised today that:

- the Millennium Development Goals (2000) gave insufficient prominence to risk reduction and resilience
- the distribution of international disaster relief is too complex, fragmented and disorganised, and needs to be properly coordinated.

Synoptic theme

Key players in seeking to reduce the burden of hazard losses are the non-governmental organisations (NGOs) – through appropriate aid and advice – and the insurance industry. But how many developing countries can afford the necessary insurance premiums?

Revision activity

Make notes summarising the ways in which people attempt to cope before, during and after one type of tectonic hazard.

Now test yourself

TESTED

12 Why is disaster aid often criticised?

Answer on p. 215

Skills reminder

You should be familiar with the following skills and techniques used in the geographical investigation of tectonic hazards:

- Analysis of global and regional distribution maps.
- Use of block diagrams to identify key features of plate boundaries.
- Analysis of time–distance maps to predict the spatial impact of tsunamis.
- Use of correlation techniques to analyse links between the magnitude of events and deaths and damage.
- Statistical analysis to compare hazard profiles.
- Interrogation of large data sets to assess data reliability and identify trends.
- Use of geographic information systems (GIS) to identify hazard risk zones and the degree of risk.

Exam practice

AS

1 (a) Name the type of plate boundary along which the most powerful earthquakes occur. **(1)**

(b) Study Figure 1.

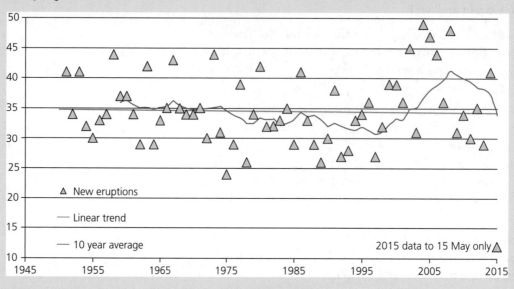

Figure 1 New volcanic eruptions per year, 1950–2015

(i) Complete Figure 1 by plotting the 33 new volcanic eruptions that occurred back in 1950. **(2)**
(ii) Describe the trends in the ten-year average. **(3)**
(c) Explain why earthquakes are more destructive than volcanic eruptions. **(4)**
(d) Explain what might be done to improve the preparedness of a community for a hazard event. **(6)**
(e) Assess the value of Park's disaster response curve. **(12)**

A-level

2 (a) Study Figure 2. Explain why plate movement is the key to understanding what the map shows. **(4)**

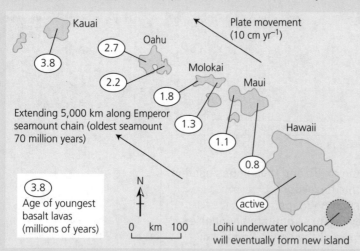

Figure 2 The ages of basalt lava flows in the Hawaiian Islands

(b) Assess the factors affecting the response to tectonic hazards. **(12)**

Answers and quick quiz 1 online

ONLINE

Exam practice answers and quick quizzes at **www.hoddereducation.co.uk/myrevisionnotes**

Summary

You should now have an understanding of:
- the global distributions and causes of earthquakes, tsunamis and volcanic eruptions
- the distinction between divergent, convergent and conservative plate boundaries
- the distributions and associated hazards of different plate boundaries
- intra-plate earthquakes and hotspot volcanoes
- the key elements of the theory of plate tectonics
- tectonic processes at different plate boundaries
- factors affecting earthquake magnitude and the type of volcanic eruption
- earthquake shock waves and secondary hazards
- volcanic emissions and secondary hazards
- factors affecting tsunamis
- the difference between a hazard and a disaster
- mega-disasters and multiple hazard zones
- disaster trends and differential impacts
- predicting and forecasting tectonic hazards
- the hazard management cycle
- Park's response curve model
- modifying tectonic events, vulnerability, resilience and losses.

2 Landscape systems, processes and change

Option A Glaciated landscapes and change

The landscapes covered in this topic may be collectively referred to as **cold environments**. They have all been affected, over considerable periods of time, by sub-zero temperatures and associated glacial and periglacial processes. The topic takes into account not only those parts of the world currently experiencing these processes but also locations that have been **glaciated** and **periglaciated** in the past.

> **Glaciation:** The modification of landscapes while covered by ice sheets or glaciers.
>
> **Periglaciation:** The modification of landscapes located adjacent to the margins of glaciers and ice sheets.

Past and present distributions of glacial and periglacial environments

> **Exam tip**
>
> You need to be aware of the distinction between areas that are currently undergoing glaciation or periglaciation and those areas that have experienced those conditions in the past but no longer do so.

Causes of climate change

REVISED

Climatic oscillations

The Quaternary period in which we live is divided into two geological epochs:
- The Pleistocene: from the beginning of the Quaternary to 11,500 years ago when the most recent continental glaciation ended.
- The Holocene: the interglacial period of today.

During the Pleistocene, the Earth's climate fluctuated between colder and warmer conditions – between **ice-house** and **greenhouse** conditions. The ice-house or glacial phases have left evidence of erosional and depositional features created by glaciers, ice sheets and their meltwaters. However, the landforms created by one glacial phase have usually been reworked, reshaped and even destroyed by later glacial phases. Today, the features produced by the most recent glacial phases are being modified by post-glacial processes.

Causes of short-term climatic oscillations

Long-term changes in the Earth's orbit around the Sun are currently seen as the primary causes of the oscillations between glacial and non-glacial conditions. The Milankovitch theory attributes the oscillations to three main characteristics of the Earth's orbit:
- Eccentricity: the orbit changes from elliptical to circular and back over a period of around 100,000 years.

- Axis tilt: this varies between 21.8° and 24.4° over a period of around 41,000 years.
- Wobble: like a spinning top, the Earth wobbles on its axis and this changes the distance from the Sun over a 21,000-year cycle.

These three different orbital cycles can combine to minimise the amount of solar energy reaching the northern hemisphere. When this happens, the climate cools and ice-house conditions return.

Evidence indicates that even within glacial and non-glacial periods, there are short-term fluctuations with frequent warming (interstadials) and cooling periods (**stadials**). Two main factors are thought to be responsible for these oscillations:

- fluctuations in the amount of energy emitted by the Sun (related to sunspot activity)
- volcanic activity – eruptions with a high **explosively index** (**VEI**) eject huge volumes of ash, sulphur dioxide, carbon dioxide and water vapour into the atmosphere. These ejected substances are distributed around the globe by high-level winds. It is thought that such eruptions can reduce the amount of solar energy reaching the Earth.

British examples of these short-term oscillations in climate are the Loch Lomond Stadial (between 12,500 and 11,500 years ago) and the Little Ice Age (between 1550 and 1750).

> **Revision activity**
>
> Just be aware of these three characteristics. It is not necessary to understand the physics behind them.

> **Typical mistake**
>
> Do not think that global warming is a recent phenomenon for which people are responsible. The history of the Earth has been one of recurring warming and cooling.

Now test yourself

TESTED

1 What is the difference between the causes of long-term and short-term climate change?

Answer on p. 215

Present and past distributions of ice cover

REVISED

> **Key concept**
>
> **Cryosphere** refers to those parts of the Earth's crust and atmosphere subject to temperatures below 0°C for at least part of each year. So this means ice sheets and glaciers, together with sea and lake ice, ground ice (permafrost) and snow cover.

Ice masses can be classified by their morphological characteristics, size and location:

- Ice sheet: complete submergence of topography beneath ice up to several kilometres deep.
- Ice cap: a smaller version of ice sheet burying upland topography.
- Ice field: not thick enough to bury an upland area.
- Valley glacier: a glacier confined within valley walls.
- Piedmont glacier: a valley glacier which spreads out beyond the valley end.
- Cirque glacier: a glacier occupying a hollow on a mountain side.
- Ice shelf: a large area of floating glacier ice extending beyond the coast.

Key concept

In terms of glacier behaviour and impact on the landscape, an important distinction is made between **warm-based** and **cold-based glaciers**:

- Warm-based glaciers, also known as 'wet' glaciers, occur in high-altitude areas outside the polar region. From the surface to the base, temperatures are close to 0°C. During the summer they generate large amounts of meltwater, which acts as a lubricant, allowing the glacier ice to slide over the bedrock.
- Cold-based glaciers, also known as 'polar' glaciers, occur in high latitudes, particularly in Antarctica and Greenland. Temperatures are well below freezing, so there is no basal sliding.

The significance of this distinction will be made evident in the sections 'Glacier systems' and 'Glacial landforms and landscapes' (pages 31–37).

Table 2.1 Estimates of present and past global ice cover

Region	Present area (estimated thousand km²)	Past area (estimated thousand km²)	Reduction in ice cover (%)
Antarctica	1350	1450	6.9
Greenland	180	235	31.9
Arctic Basin	32	1600	98.0
Andes	3	88	96.6
European Alps	0.4	4	90.0
Scandinavia	0.4	660	99.9
Asia	12	390	95.4
Rest of the world	0.2	104	99.8
TOTAL	**1578**	**4531**	**65.2**

Table 2.1 shows that the global ice cover today is one-third of what it was in the Pleistocene. Outside Antarctica and Greenland, the contraction has been so great that ice cover has almost disappeared. This applies particularly to the great mountain ranges, such as the Alps, Andes and Himalayas.

About 85 per cent of current ice cover is located in Antarctica. Next, but a long way back, comes Greenland, which accounts for just over 10 per cent. A number of factors influence the distribution of ice cover. The two most important are latitude and altitude. Other more local factors include aspect and relief.

Typical mistake

The data in Table 2.1 relate to the area covered by ice – they do not take into account the thickness of ice cover. In this respect, the volumes of the Antarctic and Greenland ice sheets may have suffered much higher percentage losses.

Now test yourself

TESTED

2 Why has there been such a great reduction in the ice cover of the major mountain ranges?
3 How does aspect affect the distribution of ice cover?

Answers on p. 215

Periglacial processes and their landforms and landscapes

REVISED

A key feature of periglacial areas is the climate:
- daily temperatures below freezing for at least nine months
- low precipitation – less than 600 mm per year
- frequent cycles of freezing and thawing
- intense frosts throughout the year.

Exam practice answers and quick quizzes at **www.hoddereducation.co.uk/myrevisionnotes**

This climate means that the ground surface is frozen for much of the year. Indeed, the ground below the surface layer is permanently frozen (permafrost). The extent of the permafrost is usually taken as indicating the distribution of periglacial environments.

> ### Key concept
> **Permafrost** is soil and rock that remain frozen as long as temperatures do not exceed 0°C in the summer months for at least two years. Three types of permafrost are recognised:
> - continuous: occurs in the coldest areas of the world where mean annual air temperatures are −6°C; it can be hundreds of metres deep
> - discontinuous: is more fragmented and less deep
> - sporadic: occurs at the margins of periglacial areas and is usually fragmented and very thin.

In summer, overlying snow and ice melt away to produce a seasonally unfrozen zone above the permafrost called the **active layer**. This may vary in thickness from a few centimetres to as much as 3 metres.

Because of the global distribution of land, most **permafrost** areas are found in the Northern hemisphere around the Arctic Ocean. Figure 2.1 shows the distribution of permafrost around the world currently.

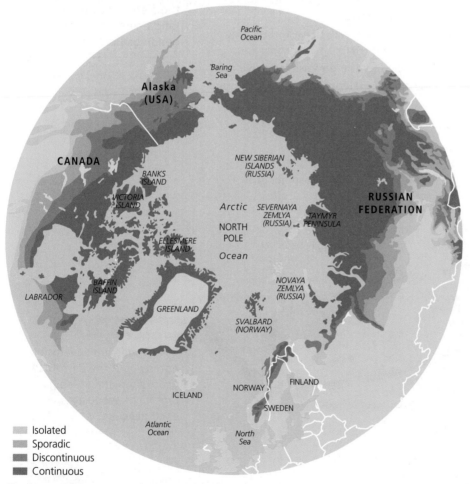

Isolated
Sporadic
Discontinuous
Continuous

Figure 2.1 Present distribution of permafrost

Periglacial processes

The distinctive climate means that periglacial areas are distinguished by a particular variety of landform processes:

- Freeze–thaw weathering: the shattering of rock as a result of water in its joints and pores freezing and expanding. Particularly active when temperatures fluctuate around freezing point.
- Frost heave: the upward movement of rock or soil particles as a result of the pressures generated by the formation of ice segregations in the ground.
- Nivation: sometimes known as snow-patch erosion, occurs when both weathering and erosion take place around and beneath a snow patch.
- Solifluction: the mass movement of the active layer downslope.
- Wind action: the lack of vegetation cover due to the low temperature plus the prevailing aridity mean there are plenty of opportunities for the strong wind to pick up and transport fine sediment. When deposited this is known as loess.
- Meltwater action: this takes place during the short summer period only, when temperatures are above freezing.

Periglacial landforms

The distinctive mix of processes is responsible for the creation of often unique periglacial landforms (Figure 2.2):

- Ground ice features: mostly caused by frost heaving and include ice-wedge polygons, patterned ground and pingos.
- Frost-shattering features: mostly caused by freeze–thaw and include block fields, scree slopes, rock glaciers and tors.
- Mass-movement features: mostly formed by the downslope movement of weathered materials and include asymmetric valleys and solifluction terraces.
- Wind action: loess plains.
- Meltwater action: braided streams.

Although periglacial conditions no longer prevail in the UK, relict periglacial features are still to be found. Most have been subject to modification by the processes of a warmer climate.

> **Exam tip**
>
> It is helpful to classify all the distinctive periglacial landforms under the five headings listed here.

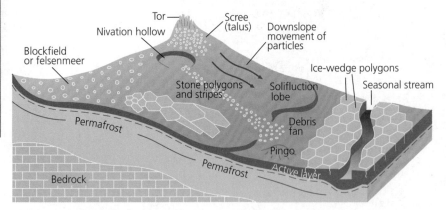

Figure 2.2 Periglacial landforms

> **Revision activity**
>
> Check that you have brief notes about the periglacial processes and at least one landform associated with each process.

Now test yourself

TESTED

4 What are the distinctive features of periglacial environments?

Answer on p. 215

Exam practice answers and quick quizzes at www.hoddereducation.co.uk/myrevisionnotes

Glacier systems

Mass balance

> **Key concept**
>
> A **systems view** of glaciers is helpful in understanding how they behave. A glacier system is made up of three components:
> - inputs: precipitation, rock debris, energy (kinetic and solar)
> - processes or throughputs: ice movement, erosion, transport, deposition
> - outputs: debris, meltwater, calving.

The glacier system has two zones:
- accumulation zone of direct snowfall, precipitation and debris avalanching from slopes above the glacier
- ablation zone, where there is a loss in the amount of ice as a result of melting, evaporation (sublimation), the calving of icebergs and ice blocks and the deposition of rock debris.

Now test yourself

5 What is the mass balance of a glacier?

Answer on p. 216

The most critical feature of the glacier system is its **mass balance**, the balance between the inputs and the outputs (Figure 2.3). Where accumulation is greater than ablation (inputs exceed outputs), a zone of excess will form and the mass balance will be positive. The glacier will grow. When the situation is reversed, with ablation greater than accumulation (output exceeding inputs), a zone of deficiency will form and the mass balance will be negative. The glacier will shrink.

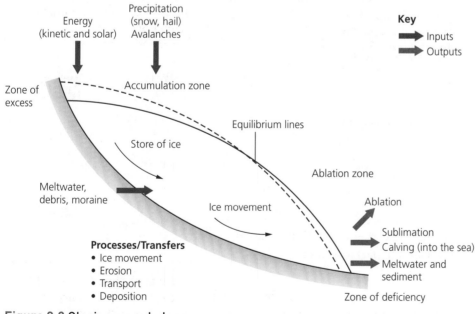

Figure 2.3 Glacier mass balance

The mass balances of glaciers vary over different time scales because of variations in the rates of accumulation and ablation. Figure 2.4 shows how the mass balance varies during the course of a year. The same principles apply when dealing with fluctuating mass balances on longer time scales, as between **stadials** and **interstadials**, or between glacials and interglacials.

> **Stadials** and **interstadials:** Short-term fluctuations within glacial periods. Stadials are colder phases that lead to ice advances, while interstadials are slightly warmer phases during which ice sheets and glaciers retreat.

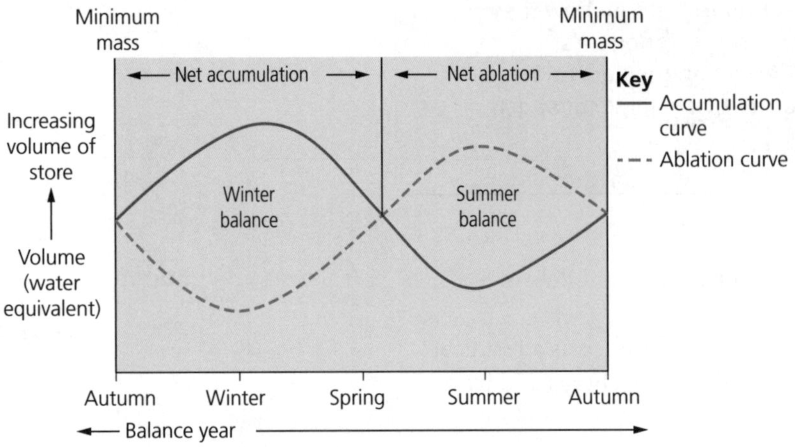

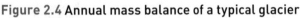

Figure 2.4 Annual mass balance of a typical glacier

Now test yourself

TESTED

6 Is the glacier a closed or an open system, and what are its components?

Answer on p. 216

Key concept

Glacier feedback comprises those effects that can either amplify a small change and make it larger (positive feedback) or diminish the change and make it smaller (negative feedback). An example of positive feedback in a glacier occurs when there is an increase in meltwater at the base of the glacier. This will accelerate basal sliding. This in turn will generate more heat from friction, thus releasing more meltwater and promoting faster sliding.

> **Exam tip**
>
> It is recommended that you have some information about fluctuating glaciers, as for example glaciers in South Georgia, the Alps or the Rockies.

Now test yourself

TESTED

7 Explain how altitude affects glacier movement.

Answer on p. 216

Glacial movement

REVISED

When it comes to glacier movement, the earlier distinction between warm-based (temperate) and cold-based (polar) glaciers becomes significant.

Warm-based glaciers move much faster than cold-based ones because of:
- the imbalance between accumulation and ablation, with the former being greater than the latter
- the availability of summer meltwater acting as a lubricant to encourage **basal sliding**

> **Basal sliding:** Occurs where ice temperatures are at or close to 0°C and a layer of basal meltwater forms between the ice and the bedrock.

Exam practice answers and quick quizzes at **www.hoddereducation.co.uk/myrevisionnotes**

- **regelation** slipping
- **internal deformation** of the basal ice.

> **Exam tip**
>
> Glacier movement is very dependent on the thawing of ice to provide meltwater, which acts as a lubricant between the glacier ice and the bedrock.

The rate of glacier movement is also controlled by other factors:

- Altitude: affects temperatures (decrease with altitude and therefore reduce speed of movement) and precipitation inputs (snow rather than rain).
- Slope: the steeper the slope, the faster the velocity.
- Lithology: friction with hard, resistant rock will tend to restrain movement at the base and sides of the glacier.
- Size: the greater the glacier mass, the greater the potential velocity.
- Mass balance: the nature of this will determine not just the velocity but also whether the glacier is retreating or advancing.

Because temperate glaciers are more mobile, they are capable of much more erosion, transportation and deposition than polar glaciers.

It is important to understand that rates of movement vary within the individual glacier, both laterally and vertically. The critical factor is friction between the glacier ice and the valley floor and sides. As a consequence, the part of the glacier moving fastest is its surface in the middle of the valley (Figure 2.5).

> **Regelation:** Occurs where basal ice is forced against a rock obstacle; it melts and then refreezes on the down-glacier side. The temporary meltwater acts as a lubricant.
>
> **Internal deformation:** A plastic-like quality caused when, under pressure, ice crystals move slightly relative to each other.

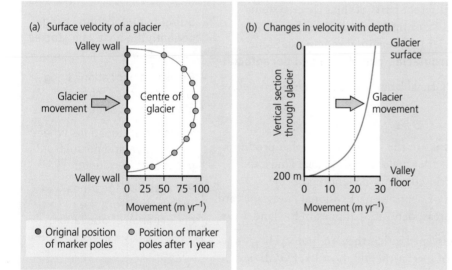

Figure 2.5 Glacier velocities

The glacier landform system

REVISED

Glacier processes

Glaciers alter landscapes and produce distinctive landforms by a number of processes.

Erosion is principally by abrasion (the scraping, scouring, rubbing and grinding action of debris being carried along by the glacier) and plucking (whereby the glacier freezes around rocks on the valley sides and floor, which are then pulled away by the movement of the glacier). A crucial aspect is entrainment, which is the incorporation of debris onto and into the glacier from subglacial and supraglacial sources.

Transport takes place at three levels: supraglacially (debris that has fallen onto the surface of the glacier), englacially (debris that has worked its way into the heart of the glacier) and subglacially (debris picked up in the basal layer from the bedrock).

Deposition occurs when material is released from the glacier ice either at its margins or at the base of the glacier. Deposition may take place directly on the ground (ice contact) or indirectly when sediments are released into meltwater and are subsequently laid down.

For more on glacial processes, see 'Glacial landforms and landscapes', below.

Glacial landforms

The distinctive glacier processes are responsible for creating a wealth of distinctive landforms. Although the same landform can often be found in many different sizes, all the glacial landforms may be broadly classified into three size groups (Table 2.2).

Table 2.2 Selected glacial landforms broadly classified by size

Macro-scale	Ice-sheet-eroded knock and lochan landscapes; cirques, arêtes and pyramidal peaks; glacial troughs and ribbon lakes; till plains, terminal moraines and sandurs
Meso-scale	Crag and tail, and roches moutonnées; drumlins, kames, eskers and kame terraces; kettle holes
Micro-scale	Glacial striations, grooves and chatter marks; erratics

These landforms are produced in different parts of the glacier system. Some are **subglacial** in origin (e.g. drumlins); others are either at or just beyond the margins of the glacier (e.g. terminal moraines). Some geographers distinguish between **proglacial**, **periglacial** and **paraglacial**.

For more on glacial landforms, see 'Glacial landforms and landscapes', below.

Glacial landscapes

Figure 2.6 shows the sequence of links in the glacier system that produces the individual landforms which together make up the glacial landscape. Those landscapes can be of a different character depending on the dominant glacial processes (erosion or deposition). Erosional landscapes are characteristically found in upland areas, depositional ones in lowland areas.

A complication with glaciated landscapes is that they are polycyclic. They are the product of several periods of glaciation and most likely will have been modified in between by periglacial or paraglacial conditions.

> **Subglacial:** locations lying beneath the base of a glacier or ice sheet.
>
> **Proglacial:** locations lying close to the ice front of a glacier or ice sheet.
>
> **Periglacial:** locations where frost action and permafrost processes dominate.
>
> **Paraglacial:** locations recovering from the disturbance of glaciation.

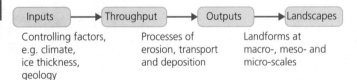

Inputs	Throughput	Outputs	Landscapes
Controlling factors, e.g. climate, ice thickness, geology	Processes of erosion, transport and deposition	Landforms at macro-, meso- and micro-scales	

Figure 2.6 The glacier landscape system

Now test yourself

TESTED

8 What glaciated landforms would distinguish an upland glaciated landscape from a lowland one?

Answer on p. 216

Glacial landforms and landscapes

Glacial erosion

REVISED

Processes

The rate, intensity and effectiveness of glacial erosion vary with time and from place to place within the world's cold environments. Figure 2.7 summarises the main factors influencing the rates of **abrasion** and **plucking**.

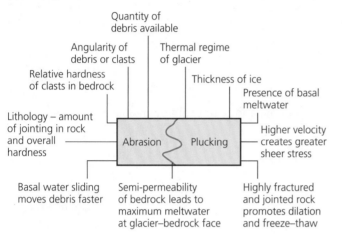

Figure 2.7 Factors affecting abrasion and plucking rates

Macro landforms

Macro features include U-shaped valleys with their truncated spurs and tributary hanging valleys. Also conspicuous are the three related landforms of cirques, arêtes and pyramidal peaks.

Meso landforms

Some meso landforms, such as roches moutonnées and whalebacks, are found within macro features such as glacial troughs. However, meso landforms are more commonly found where ice sheets and glaciers spread out over large areas of lower relief. They are largely the product of ice scouring.

> **Abrasion:** The process by which solid rock is eroded by rock fragments being transported by glaciers.
>
> **Plucking:** The detachment of joint-bounded blocks by glaciers. It was thought that the ice froze onto the rock and wrenched blocks of it away. However, it is now believed that the breaking away of blocks is due to the movement of the glacier.

> **Revision activity**
>
> Make brief notes about three factors affecting each of the two processes.

> **Exam tip**
>
> Remember that there are other, less significant erosional processes at work. These are quarrying, crushing, basal melting, freeze–thaw and mass movement.

> ### Now test yourself
> TESTED
>
> 9 How are cirques, arêtes and pyramidal peaks related?
>
> Answer on p. 216

Glacial deposition

REVISED

All accumulations of glacial debris are referred to as **moraines**. There are two broad types:

- Subglacial: moraines deposited beneath the glacier. These are made up largely of lodgement till and create extensive flat areas that cover pre-existing topography. In places, the till is moulded into elongated, streamlined mounds (drumlins).

- Ice-marginal: moraines deposited along the edges of glaciers. They include lateral moraine (along the edge of the valley floor), medial moraine (along the middle of the valley floor), terminal moraine (a ridge of moraine extending across the valley at the furthest point reached by the glacier) and recessional moraine (a series of cross-valley ridges behind the terminal moraine).

> **Exam tip**
>
> Be aware that not all drumlins are thought to have been formed in the same way. There are two main theories, one emphasising the deposition of moraine in the lee of a solid obstacle and the other the impact of subglacial meltwater.

Now test yourself

TESTED

10 Distinguish between a) terminal and recessional moraines, and b) lateral and medial moraines.

Answer on p. 216

Glacial meltwater

REVISED

Meltwater from glaciers plays a vital role in the fluvio-glacial processes of erosion, entrainment, transport and deposition. The two main sources of meltwater are:

- surface melting produced mainly during the short summer
- basal melting caused by the temperature of the ice being at pressure melting point. Meltwater beneath a glacier facilitates basal sliding, subglacial bed formation and erosion.

> **Key concept**
>
> **Pressure melting point** is the temperature at which ice melts at a given pressure. It is normally 0°C at the surface of a glacier, but inside the melting point will be fractionally lowered by the raised pressure within the ice.

Fluvio-glacial deposits differ from glacial ones in that they are rounded rather than angular, sorted by size, with the heaviest load deposited first – glacial unsorted – and stratified into distinct layers by the meltwater – glacial unstratified.

The main features resulting from subglacial meltwater are:

- eskers: long sinuous ridges of the valley floor
- kames: mounds of stratified material deposited beneath the ice of a decaying glacier or ice sheet
- kame terraces: ridges of material deposited along the edge of a valley floor.

The main proglacial features created by meltwater as it exits the snout of a glacier are:

- outwash plains (sandurs): formed when meltwater spreads out after issuing from the glacier snout and its load is deposited to create a gently sloping surface

- proglacial lakes: created when meltwater becomes impounded between the glacier snout and high ground
- overflow channels: spillways cut when proglacial lakes overflow their confines.

Now test yourself

TESTED

11 What is the significance of the pressure melting point?

Answer on p. 216

The use and management of glaciated and glacial landscapes

Intrinsic value

REVISED

Relict glaciated and active glacial and periglacial landscapes are valued by people for a variety of reasons. For their:
- **wilderness** and the opportunity for people to escape from the hustle and bustle of modern living
- spectacular scenery
- unique biodiversity, ecology and wildlife
- challenge to scientists, explorers and travellers
- role in the global water and carbon cycles
- glaciers – an important source of water
- economic opportunities – farming, forestry, tourism, mining, quarrying and hydroelectric power (HEP)
- artistic inspiration to generations of artists and writers
- cultural heritage created by indigenous peoples.

> **Key concept**
>
> **Wilderness** is any area of the world that has remained relatively untouched by human activity (for example, Antarctica), but may be home to small numbers of indigenous people (for example, Amazonia). The overriding characteristics of wilderness are remoteness and detachment from the pace and pressures of modern living. The value of wilderness can be spiritual, scientific, artistic and therapeutic. But what now most threatens wilderness are its economic potential as a source of resources and its appeal as a tourist destination.

> **Exam tip**
>
> Be aware of the distinction between present and past glaciated landscapes. The former are rather more in focus in this part of the topic.

Clearly, the value of wilderness is much greater than an economic one. The quality of wilderness offered by these landscapes varies from place to place and is inversely proportional to the degree to which they have been modified by people (Figure 2.8). The **wilderness continuum** runs from the almost pristine and active Antarctic and Arctic to the well trampled and commercially exploited relict landscapes of Snowdonia and the Cairngorms.

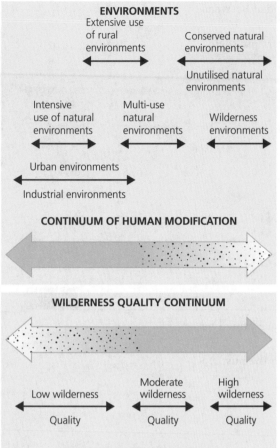

Figure 2.8 The wilderness continuum

Synoptic theme

It is unfortunate that not all the players involved in glacial and periglacial environments share the same attitudes. What they seek ranges from the outright exploitation of resources to resource preservation.

Threats

REVISED

The threats facing active and relict glaciated landscapes fall into three categories:

- the natural hazards that pose a risk to humans
- the impacts of human activity
- global warming.

Natural hazards

It is only the human presence in these environments that converts the following natural events into possible hazards:

- Avalanche: a rapid descent of a large mass of rock, ice and snow down a mountain slope. Avalanches are frequently triggered by human activity, such as ski-ing.
- Lahar: a rapid flow of mud and debris resulting from meltwater overflowing a glacial lake.
- Glacial outburst flood (jökulhlaup): a short-lived but sometimes catastrophic flood resulting from the sudden release of meltwater stored within or on the surface of a glacier or ice sheet. The build-up of meltwater is often triggered by the escape of geothermal heat beneath the ice.

Human impacts

Clearly, it is human activity that truly threatens to degrade and damage fragile glaciated landscapes. The following activities may be singled out as causing the most concern:

- Polar tourism: a booming but very expensive business. To date it is fairly well policed by the International Association of Antarctica Tour Operators (IAATO).
- Mountain tourism: the modern tourist infrastructure and the sheer pressure of tourist numbers are inflicting considerable environmental damage. In particular, there is the wear and tear on fragile ecosystems, soil erosion and upsetting water budgets that are critical to the supply of water to rivers and nearby populated areas.
- Mining and quarrying: possibly the biggest threat here comes from the oil industry. The Alaskan experience serves as a reminder of the immense damage that this industry can inflict on the environment – roads and pipelines cutting swathes across the tundra, pollution around oil wells and oil spills in coastal waters.

> **Synoptic theme**
>
> The very presence of people, let alone economic players, has a debilitating impact on the resilience of fragile landscapes.

One particular threat concerns the indigenous peoples of the northern hemisphere who still make their living in these environments, be it by fishing and sealing or herding caribou. It is important that they, their traditional livelihoods and cultural heritage be protected.

> **Synoptic theme**
>
> It is often the case that the actions of players have indirect and unforeseen impacts on the natural systems of glaciated landscapes. For example, IAATO, by encouraging responsible polar tourism, is also encouraging more people to visit Antarctica. The means more people damaging fragile natural systems.

Global warming

This is undoubtedly the greatest threat to glacial environments. Already most of the world's glaciers are in retreat. Data from satellite surveys of Greenland show a huge decrease in the ice-covered area. Surveys of Antarctica are showing alarming losses of shelf ice.

These changes will have serious consequences for the global population: the disruption of global water and carbon cycles, as well as a rise in sea level that threatens coastal cities and millions of inhabitants.

Now test yourself

TESTED

12 Which of the human impacts on glaciated environments poses the greatest threat? Give your reasons.

Answer on p. 216

Management approaches

There are various stakeholders with an interest in the future of glaciated landscapes. They range from conservationists to oil companies, from scientists to tourist operators, from indigenous people to investors.

The way ahead

There are a number of possible approaches to the management of cold environments (Figure 2.9). They range from 'do nothing' and 'business as usual' to 'comprehensive conservation' and 'total protection'.

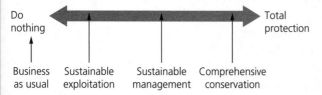

Figure 2.9 The spectrum of management options for cold environments

Which strategy is most appropriate depends on the area and the views of the players involved. In the event that unanimity will rarely prevail among stakeholders with different values and viewpoints, compromise will be the only way ahead. Possible strategies here include:

● sustainable exploitation
● sustainable management
● zoning – a useful middle way, with the high-value environments given full protection and buffer zones created between these areas and zones where sustainable activities are permitted.

> **Synoptic theme**
>
> The attitudes and actions of players in glaciated landscapes range from all-out exploitation to preservation.

Legislation

If it is to be successful, any management of glacial and glaciated environments requires the support of legislation, both international and national.

Examples on the international scale include the Antarctic Treaty (1959), which established that all areas south of 60°S were to be without any national claims and protected for peace and science. Since then, a large number of recommendations and international agreements have been adopted to form the Antarctic Treaty System (ATS). While the Antarctic now looks reasonably 'future proof', the same cannot be said for the Arctic, an oceanic area largely surrounded by the two superpowers of Russia and the USA.

Effective legislation on the national scale has led to the creation of a range of conservation areas, from national parks to more local nature reserves, particularly in parts of the Arctic and in high mountainous regions such as the Alps, Himalayas and Rockies.

The problem with well-intentioned national legislation is the lack of overarching international legislation. Without it, conservation and protection become piecemeal. The crucial point is that the future of the world's cold environments is a global issue requiring the backing of international treaties to give real 'teeth' to both conservation and protection.

> **Revision activity**
>
> Make a list of these stakeholders and identify their particular visions of glaciated landscapes.

> **Revision activity**
>
> Find out more about why and how global warming, resource exploitation and geopolitics are threatening the future of the Arctic.

Exam practice answers and quick quizzes at **www.hoddereducation.co.uk/myrevisionnotes**

Climate change

Much attention is focused today on carbon emissions as a causal factor in climate change and on the need for concerted international actions to reduce these emissions. However, what if today's global warming is simply one of those natural changes in climate that have characterised the Earth during both geological and historic times? If this is the case, then there is little that humans can do, on any spatial scale, to change the forces and courses of nature.

Revision activity

For further information relating to the issue of climate change, see Topic 6.

Now test yourself

TESTED

13 What are the options when it comes to the possible management of cold environments?

Answer on p. 216

Synoptic theme

The risks associated with climate change are creating **uncertainty**. This is creating a need to devise appropriate mitigation and adaptation strategies.

Skills reminder

You should be familiar with the geographical skills used in the investigation of the following aspects of glaciated landscapes:

- representing reconstructed past climates by means of graphs
- comparing past and present distributions of glacial and glaciated landscapes using global and regional maps
- calculating the mass balance and the position of the equilibrium line using data from global and regional maps
- identifying the main features of glaciers using GIS
- comparing rates of glacier movement using measures of central tendency
- analysing and correlating cirque orientation, height and size using data from large-scale maps
- representing till fabric analysis by means of rose diagrams
- reconstructing past ice extent and ice flow direction from Ordnance Survey and British Geological Survey (drift) maps.
- analysing changes in sediment size and shape in glacial and fluvio-glacial deposits using the student's t-test and measures of central tendency
- analysing data relating to mean rates of glacial recession in different regions
- comparing drumlin morphometry and orientation in two contrasting locations using field measurements.

Exam practice

AS

1 (a) State **one** erosion process that occurs in glaciated upland areas. (1)

(b) Study Figure 1.

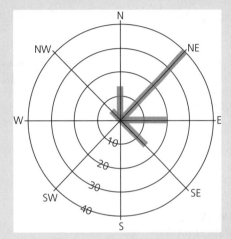

Figure 1 The orientation of cirques in the Lake District

(i) What is the least common cirque orientation? (1)
(ii) Approximately how many of the cirques faced this direction? (1)
(iii) Explain the general orientation of the cirques. (3)

(c) Explain the formation of lateral moraines. (4)

(d) Explain the main processes of periglaciation. (6)

(e) Assess the need to manage glaciated environments. (12)

A-level

2 (a) Study Figure 2. Explain the occurrence of high erosional intensity. (6)

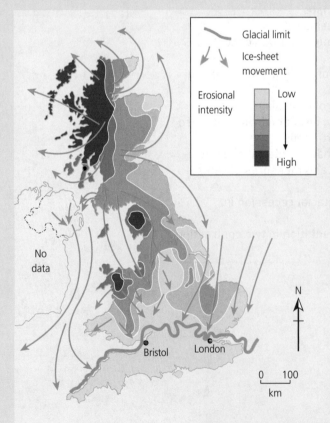

Figure 2 Ice sheet movement and erosional intensity in the UK during the last glacial advance

Exam practice answers and quick quizzes at **www.hoddereducation.co.uk/myrevisionnotes**

(b) Explain how meltwater contributes to the glaciated landscape. (6)

(c) Explain how the balance between net accumulation and net ablation is the key to understanding glacier behaviour. (8)

(d) Evaluate the extent to which glacial landscapes are more threatened by climate change than periglacial landscapes. (20)

Answers and quick quiz 2A online

ONLINE

Summary

You should now have an understanding of:

● the causes of long- and short-term changes in climate and the creation of icehouse and greenhouse conditions

● the present distribution of ice cover and its role in global systems

● evidence for the extent of ice cover during the Pleistocene

● periglacial processes and their distinctive landscapes

● glacial dynamics and glacial systems at work

● glacial movement and variations in its rates

● the glacier landform system

● glacial erosion and its contribution to the glaciated landscape

● glacial deposition and its contribution to the glaciated landscape

● glacial meltwater and its contribution to the glaciated landscape

● the value of glacial and periglacial landscapes

● the threats facing glaciated landscapes

● managing the threats to glaciated landscapes.

Option B Coastal landscapes and change

The coast is a very important component of the globe. Many of the world's major cities are in coastal locations so large numbers of people live and work there. The coast is a critical interface between land and sea, subject to both terrestrial and marine processes, and it experiences extreme events, such as tropical cyclones, tsunamis and storm surges.

Different coastal landscapes and their processes

The coast and littoral zone

REVISED

All coastlines show the same littoral sub–zones (Figure 2.10), but not all coastlines have similar landscapes. The **littoral zone** can be a dynamic one of rapid change.

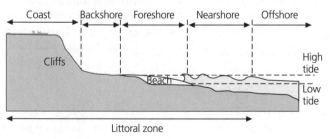

Figure 2.10 The littoral zone and its sub-zones

> **Littoral zone:** The wider coastal zone, which includes adjacent land areas, the shore and the shallow part of the sea just offshore. It comprises four sub-zones: coast, backshore, foreshore and nearshore (Figure 2.10).

Classifying coasts

Coasts may be classified in various ways, as for example by:
- geological characteristics (lithology and structure)
- the impacts of sea-level changes (rising or falling)
- the dominant coastal process (erosion or deposition).

A common but simple coastal classification distinguishes between:
- rocky or cliffed: where there is a clear distinction between land and sea, mainly because of the height of the cliffs. Exposure to the erosive forces of the sea, rain and wind creates a high-energy coastline.
- coastal plains: where the land slopes gently towards the sea and there is an almost imperceptible transition from one to the other. These are often maintained in a state of **dynamic equilibrium** between the deposition of sediment by river stems entering the sea and sediment from offshore sources and marine erosion. They are typically low-energy coastlines.

Rocky coasts and coastal plains

Rocky coasts result from a geology that is resistant to the erosive forces of the sea and weather in high-energy environments. Coastal plains are found in areas of low relief and depend on the supply of terrestrial sediment.

> **Dynamic equilibrium:** The balanced state of a system when inputs and outputs balance over time. If one of the inputs changes, then the internal equilibrium of the system is upset. By a process of feedback, the system adjusts to the change and the equilibrium is regained.

> **Exam tip**
>
> Be sure you are able to name actual stretches of coastline that exemplify both cliffed and coastal plain types. These will help add authenticity to your answers.

Now test yourself

TESTED

14 What is the significance of the difference between a high-energy and a low-energy coastline?

Answer on p. 216

Geological structure and the development of coastal landscapes

REVISED

> **Key concept**
>
> **Geological structure** refers to the arrangement of rocks in three dimensions and involves three key elements:
> - strata: the different layers of rock in a location and how they relate to each other
> - deformation: the degree to which rock strata have been tilted or folded by tectonic activity
> - faulting: the presence of fractures along which rocks have moved.
>
> All three elements affect coastal landscapes and the development of coastal landforms.

> **Typical mistake**
>
> Do not confuse geological structure and lithology. The former is about the disposition of rocks, while the latter is about the composition of rocks, whether they are limestones, sandstones, marls, etc.

Concordant and discordant coasts

Geological structure produces two main types of coast (Figure 2.11):
- concordant: formed when rock strata run parallel to the coastline. The typical coastline is generally a smooth or slightly indented one, also known as Dalmatian type.
- discordant: formed when different rock strata intersect the coast at an angle, so that lithology varies along the coastline. The typical coastline is one of bays and headlands, also known as Atlantic type.

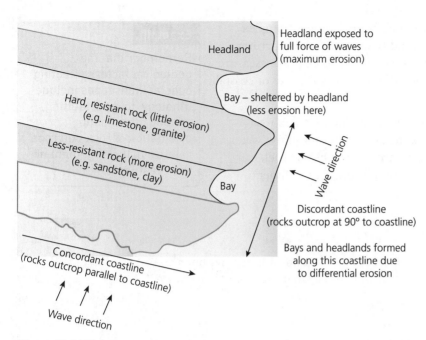

Figure 2.11 **The impacts of lithology and rock structure on coastlines**

Cliff profiles

Cliff profiles are influenced by two different aspects of geology: the resistance of the rock to erosion and the dip or angle of rock strata in relation to the coastline.

In addition to the dip of strata, other geological features influence cliff profiles and rates of erosion:

- faults: rocks are fractured and therefore weakened on either side of a fault line
- joints: occur in most rocks and are potential lines of weakness
- fissures: small cracks in rocks also represent weakness that erosion can exploit.

Exam tip

The Isle of Purbeck (Dorset) is part of the Jurassic Coast and provides excellent examples of the impact of geology on the coast landscape.

Rates of coastal recession

REVISED

Geological factors

The rate of coastal erosion and recession is a critical aspect of the coastal zone. The rate is influenced by many factors but the most significant is lithology (rock type). The three major rock types erode at different rates (Table 2.3).

Table 2.3 **Rates of erosion for different rock types**

Rock type	Examples	Erosion rate
Igneous	Granite Basalt Dolerite	Very slow Rocks are resistant to erosion because they are crystalline and usually have few joints
Metamorphic	Slate Schist Marble	Slow Crystalline rocks are resistant to erosion Metamorphic rocks are often folded and fractured and therefore vulnerable to erosion
Sedimentary	Sandstone Limestone Shale	Moderate to fast Most sedimentary rocks erode faster than the other two types Younger rocks tend to be softer and weaker Rocks with bedding planes and fractures are more vulnerable to erosion

Differential erosion of alternating strata in cliffs can produce complex cliff profiles. The lithology of the strata and their geological structures, together with their degree of exposure to the erosive forces of the sea, are the major factors influencing the rate of cliff retreat and the rate of coastal recession.

Permeability is one other geological factor that should be taken into account. Permeable rocks, such as sandstone and limestone, allow water to pass through them. Groundwater flow through permeable rock can weaken rocks by removing the cement that binds the rock sediment. Slumping is a common outcome.

Now test yourself

TESTED

15 What distinguishes metamorphic rock from igneous and sedimentary ones?
16 How is the resistance of rocks to erosion affected by their geological structure?

Answers on p. 216

Coastal vegetation

Vegetation is an important factor influencing the coastal landscape. Many coastlines are protected from the erosion of unconsolidated sediment by the stabilising influence of plants, as for example in coastal sand dunes, salt marshes (haloseres) and mangrove swamps.

Many of the plants that grow in coastal environments are **halophytes**; some are **xerophytes**.

Sand dunes are particularly effective in encouraging coastal accretion. Through **plant succession**, sand dunes can convert a supply of sediment into land. The succession starts with specialised halophytic plants capable of growing in salty, bare sand (Figure 2.12). Once established, they trap more sand and this leads to the formation of embryo dunes. The embryo dunes, in their turn, alter the environmental conditions to an environment in which xerophytic plants can flourish. So the succession continues. The dunes gradually become fixed and the plant cover develops into a climax community of heath or woodland.

Halophytes: Plants that can tolerate salt water, be it around their roots, being submerged at high tide or being sprayed by the sea.

Xerophytes: Plants that can tolerate very dry conditions, such as those found in coastal sand dunes.

Plant succession: The sequential development of vegetation from its initial establishment on bare ground through to the ultimate vegetation cover or climax plant community.

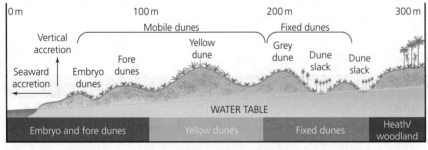

Figure 2.12 Cross section across sand dunes showing plant succession

A similar process of plant succession occurs on bare mud deposited in estuaries. Estuaries are ideal for the development of salt marshes because of the sheltered conditions and the supply of mud and silt provided by the river. The succession starts with algae, followed by various halophytic grasses, then sea thrift and lavender and ending with a climax community of rush and sedge.

Now test yourself

TESTED

17 Explain why wind is important in the formation of sand dunes.

Answer on p. 216

Coast landforms and landscapes

Marine erosion

REVISED

Waves

Waves are caused by friction between wind and water. They are extremely important along any stretch of the coast as they directly influence the three marine processes of erosion, transport and deposition. Wave size and strength depends on:

- the strength of the wind
- the length of time the wind blows for
- water depth
- wave **fetch**.

> **Fetch:** The uninterrupted distance across water over which the wind blows. It is the distance over which waves are able to grow in size.

> **Typical mistake**
>
> Waves and tides are frequently confused. Tides are formed by the gravitational pull of the Moon on water on the Earth's surface. This causes sea level (the tide) to rise and fall twice a day. Tidal ranges vary from place to place. Waves are more localised disturbances of the sea caused mainly by wind.

Now test yourself

TESTED

18 What is happening to the water particles within a wave?

Answer on p. 216

Two types of wave are distinguished (Figure 2.13). **Constructive waves** are of low height and long length. They are sometimes described as 'spilling' waves with a strong **swash** and a weak **backwash**. The strong swash pushes sediment up the beach and deposits it in a ridge at the top of the beach. The weak backwash means that the deposited material is not dragged back to the sea. **Destructive waves** are relatively high with a short length. This makes them 'plunging' waves with a strong backwash that erodes and carries away beach material.

> **Swash:** The flow of seawater up a beach as a wave breaks.
>
> **Backwash:** The return flow of seawater back down the beach to meet the next incoming wave.

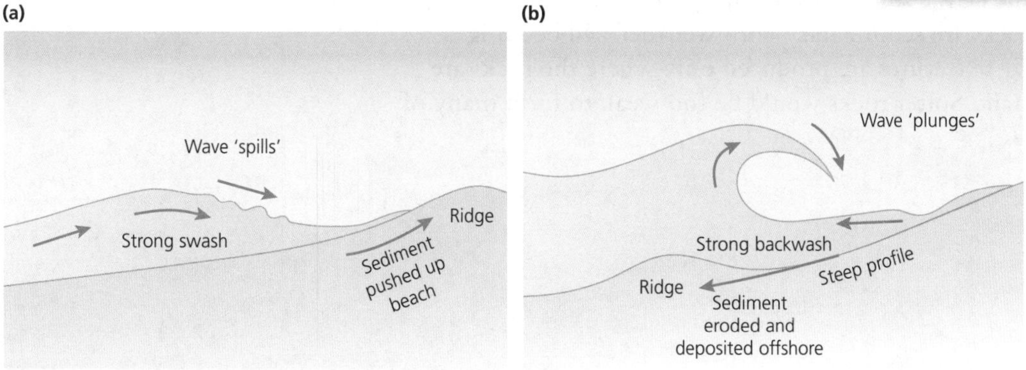

(a)

(b)

Wave 'spills'

Strong swash

Sediment pushed up beach

Ridge

Wave 'plunges'

Strong backwash

Steep profile

Ridge

Sediment eroded and deposited offshore

Figure 2.13 Constructive (a) and destructive (b) waves

Beach morphology is strongly conditioned by the nature of the prevailing waves. But wave conditions can fluctuate over time and bring with them changes to beach morphology. There are four diagnostic beach features of prevailing wave conditions:

- storm beach: the result of constructive waves during stormy weather
- berms: small ridges built by constructive waves during relatively calm weather
- cusps: the product of gentle destructive waves eroding berms
- offshore bars: formed by persistent destructive waves.

> **Beach morphology:** The shape of a beach, including its width and slope (the beach profile) and features such as berms, ridges and runnels. It also includes the type of sediment (shingle, sand, mud) forming the beach.

Now test yourself

TESTED

19 Identify the sources of beach material.

Answer on p. 216

Erosion processes

Waves cause erosion. Erosion is not necessarily a continuous process. It mostly occurs during storms and when:

- waves approach the coast at right angles
- the tide is high
- heavy rainfall has weakened the rocks of the cliff
- debris at the foot of the cliff has been removed and no longer protects this critical point.

Now test yourself

TESTED

20 Why does most erosion take place during storms?

Answer on p. 216

There are four main types of marine erosion:

- hydraulic action: wave quarrying when air trapped in joints and cracks is compressed by the force of waves crashing against the cliff
- abrasion (corrasion): sediment being carried in the waves has a wearing-down effect
- attrition: the wearing down of sediment as it is moved around by the waves
- corrosion (solution): carbonate rocks such as limestone are dissolved by rainwater, sea spray and seawater.

Erosional landforms

Figure 2.14 shows the classic suite of landforms produced by marine erosion where the rocks are sedimentary with well-defined bedding planes and joints. These features are produced only where the rocks are fairly hard and resistant. Softer rocks would be too weak to form many of these erosive features.

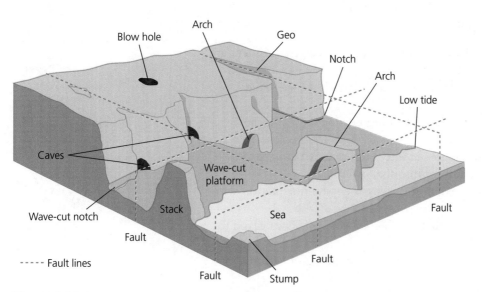

Figure 2.14 **Erosional landforms in resistant sedimentary rocks**

Perhaps the most critical erosional feature is the wave-cut notch formed by the processes of hydraulic action and abrasion. As the notch becomes deeper, the rocks overhanging it become unstable and eventually collapse. Repeated cycles of notch cutting and collapse cause cliffs to recede inland.

Revision activity

Be sure you know a location where the features shown in Figure 2.14 occur.

Now test yourself

TESTED

21 Explain the sequence of erosional landforms that starts with a cave and finishes with a stack.

Answer on p. 216

Marine transport and deposition

REVISED

Sediment transportation

Sediment is transported by the sea in four different ways:

- traction: heavier sediment (pebbles, boulders) rolls along the sea floor, pushed by waves and **currents**
- saltation: sediment (mainly sand particles) bounces along the floor
- suspension: fine sediment (silt, clay) is carried within the body of water
- solution: dissolved sediment (calcium carbonate) is carried in the water as a solution.

Much of the transport of sediment takes place along the coast rather than into and away from the shore. This is known as **longshore drift**. Such a movement is produced by waves approaching the coast at an angle. The swash pushes sediment obliquely up the beach and backwash returns it directly to the sea. The combination results in sediment slowly moving along the coast in the same general direction as the waves approach the shore.

Currents: Flows of seawater in a particular direction driven by wind, tides and differences in density, salinity and temperature.

Exam tip

Remember that wave direction is determined by the direction of the wind. Because of this, the general direction of longshore drift reflects the direction of the prevailing wind.

Now test yourself

TESTED

22 Is longshore drift a current? Give your reasons.

Answer on p. 216

Depositional features

The main features produced by the deposition of transported sediment are (Figure 2.15):

- bayhead beach: an accumulation of sand at the head of a sheltered stretch of water between two headlands
- spit: a sand or shingle beach ridge extending beyond a turn in the coastline
- recurved hooked spit: a spit built out into a bay or across an estuary, the end of which curves landward into shallower water
- bar: a sand or shingle beach extending across a coastal indentation with a lagoon behind
- tombolo: a sand or shingle bar that attaches a former offshore island to the coast
- cuspate foreland: a triangular area of shingle extending out from a shoreline, possibly formed by longshore drifts from opposing directions.

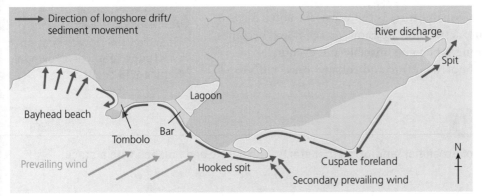

Figure 2.15 Depositional landforms of the coast

Sediment cells

Recognising the existence of **sediment cells** and understanding how they work is of fundamental importance to the management of the coast, particularly at a time of global warming and when there is much human pressure on the coast.

> **Revision activity**
>
> Make notes about one of the sediment cells around the UK's coast, preferably one on a stretch of the coast that you know about firsthand.

Sub-aerial processes

REVISED ☐

Weathering and **mass movement** are processes that affect most coastlines.

Weathering

> **Typical mistake**
>
> Do not confuse weathering with erosion. Weathering does not involve any form of movement whereas erosion does. This is due to the action of external forces, such as wind and running water.

Sediment cells: Long stretches of coastline that operate as almost self-contained physical systems. In each sediment cell there are sources (inputs), where sediment is generated (e.g. eroding cliffs and beaches), transfer zones (flows), which are stretches where sediment is moving along the coast by longshore drift and currents, and sinks (outputs), locations where the dominant process is deposition (e.g. spits and offshore bars).

Weathering: The disintegration and decomposition of rocks in situ by the combined actions of the weather, plants and animals.

Mass movement: A collective term for the processes responsible for the downslope movement of weathered material under the influence of gravity.

There are three types of weathering:
- mechanical: the breakdown of rock by some form of physical force
- chemical: involving a chemical reaction and decomposition
- biological: the actions of bacteria, plants and animals, which speed up mechanical or physical weathering.

The processes of weathering at work in any location are conditioned by the climate (rather than the weather). Temperature and precipitation are the key climatic elements.

Now test yourself

TESTED

23 What type of weathering prevails in the hot, humid parts of the world?
24 Where in the world is mechanical weathering most effective, and why?

Answers on p. 216

Mass movement

Mass movement can be classified in a number of ways, such as by the speed of movement and the type of material moving (solid rock, debris or soil). The following are types of mass movement common along cliffed coastlines:
- Rockfalls occur where the rock on a cliff being undercut by the sea is weakened by weathering. Falls can be sudden and spectacular.
- Rotational slides are slow downslope movements of a mass of rock or debris over a curved plane. They are particularly common where a permeable rock (e.g. sandstone) overlies an unstable impermeable rock (e.g. clay). Water passing through the permeable rock lubricates the junction with the underlying impermeable rock, thus facilitating the sliding.
- Landslides are sudden downslope surges occurring when weathered rock and soil become saturated and lubricated by water (meltwater, springs).

The distinctive coastal landforms created by mass movement include screes (the product of mechanical weathering of an exposed rock face), rotational scars (the outcome of slumping) and cliff terraces (the result of rotational sliding).

Now test yourself

TESTED

25 What is the difference between a) rockfalls and rotational slides, b) screes and cliff terraces?

Answer on p. 216

Coastal risks

Sea-level change

REVISED

Sea levels change on a day-to-day basis as a result of tides, changes in atmospheric air pressure and winds. Such changes are very short term. However, over long time scales, sea-level changes are more permanent and the outcome of complex factors. Part of the complexity is that a change in sea level can be brought about by a change either in land level (**isostatic change**) or in the volume of the sea (**eustatic change**).

Long-term changes

A marine regression results from a eustatic fall in sea level (as during glacial periods when water becomes locked up in ice and snow) and an isostatic fall in sea level (when ice sheets melt and the land rises). Both movements expose the seabed and produce an emergent coast (Figure 2.16).

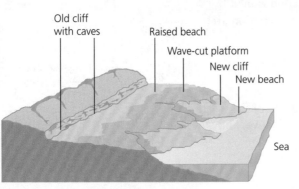

Figure 2.16 Features of an emergent coastline

Emergent and submergent coastlines

A marine transgression results from a eustatic rise in sea level (at the end of a glacial period) and an isostatic rise in sea level (when land sinks under the weight of accumulated snow and ice). In both cases, large areas of land are submerged beneath the sea, producing a submergent coast. The ria coastline is one of the best examples of such a coast (Figure 2.17).

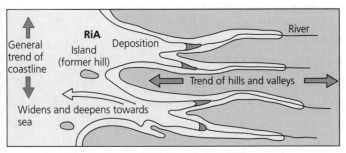

Figure 2.17 A ria or submergent coastline

Now test yourself

TESTED

26 What are barrier islands, and how are they formed?

Answer on p. 217

Contemporary sea-level change

Sea levels are rising globally and most scientists attribute this to the impact of global warming. The current rate of rise is about 2 mm per year. There are two components:

● thermal expansion of the oceans as they are warmed by the change in climate
● the melting of ice sheets and glaciers increasing the water volume of the oceans.

In some locations, shorelines are being either lifted or dropped by earthquakes. For example, the earthquake responsible for the 2004 Indian Ocean tsunami caused the coastline of part of Sumatra to drop by 1 metre, while some offshore islands were raised by up to 2 metres.

Coastal recession

Rapid recession

Rapid coastal recession threatens people, their property and their livelihoods. It is caused by physical factors, such as:
- long wave fetch and large destructive waves
- strong longshore drift
- soft or unconsolidated geology
- cliffs with structural weaknesses and vulnerable to weathering and mass movement.

It can also be accelerated by human actions, such as:
- dredging the offshore seabed for sand and gravel
- river dams reducing the supply of sediment to the coast (e.g. the impact of the Aswan High Dam on the Nile Delta)
- coastal management (construction of groynes).

Coastal erosion and recession are not constant, even on coasts, such as the Holderness coast of Yorkshire and that of California, which are receding rapidly. Changes in the weather account for much of this variation in the rate of erosion.

> **Synoptic theme**
>
> The actions of players may alter the natural systems of the coast, either deliberately (by reclaiming coastal marshes and mudflats) or inadvertently (by constructing groynes to protect a stretch of coastline and so disrupting longshore drift).

> **Typical mistake**
>
> It is wrong to attribute the present rapid erosion of some coastlines to climate change. In most locations, erosion has been going on for centuries, even millennia. Remember that climate change will simply accelerate what is already under way.

Now test yourself

27 How is it that river dams constructed a long way from the coast can contribute to coastal erosion?

Answer on p. 217

Coastal flooding

Factors

Coastal flooding is another significant and increasing risk along some low-lying coastlines. High-risk areas include coastal plains, estuaries and deltas. The risk is being increased by:
- the rising sea level associated with global warming
- human actions, such as the removal of coastal vegetation (e.g. mangroves), the building of coastal tourist resorts and the general pressure of population drawn, for example, to deltas by the fertile soils and farm productivity. Asia's six mega deltas are home to more than 100 million 'at risk' people.

Storm surges, short-term rises in sea level caused by low air pressure, are a particular hazard in low-lying coastal areas and are capable of causing immense damage and destruction. Classic examples include the North Sea storm surges of 1953 and 2013 (Figure 2.18).

> **Revision activity**
>
> Be sure to have some examples of islands (e.g. the Maldives, Tuvalu and Vanuatu) and deltaic areas (e.g. Bangladesh) that are particularly exposed to coastal flooding.

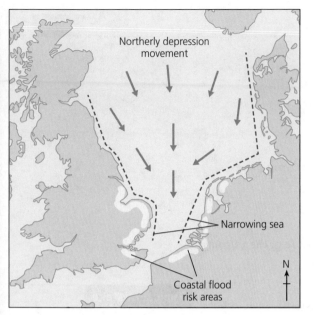

Figure 2.18 North Sea coastal topography and storm surges

Climate change

It is important to remember that storm surges, cyclones and depressions have always been a part of weather in many locations around the world. No doubt they will continue to be hazards, but the concern is that the frequency and magnitude of such events will be intensified by global warming. In short, many believe that coastal flooding will become still more of a threat to human lives and livelihoods.

> **Synoptic theme**
>
> The risks associated with global warming are resulting in uncertainty. This is creating a need to devise appropriate mitigation and adaptation strategies.

Now test yourself   TESTED

28 What causes storm surges?

Answer on p. 217

Coastal management

Risks and their consequences for communities   REVISED

The risks of coastal recession and coastal flooding look set to increase in the near future. These will clearly affect coastal communities. The costs of both changes fall into three broad categories:
- Economic costs: these include the loss of property in the form of homes and businesses, the loss of transport lines, as well as the loss of farmland and other means of livelihood.

- Social costs: these include the impacts on people, such as the costs of relocation and community disruption, as well as the impacts on health and well-being, levels of stress and hardship.
- Environmental costs: the loss of coastal habitats and ecosystems.

Of course, the higher the population density and the greater the concentration of economic wealth, the greater the consequences and the higher the losses involved.

While in many parts of the world rising sea levels will be managed by building higher flood walls and stronger sea defences, there will be locations where the situation becomes unmanageable. Most at risk are islands such as the Maldives, Tuvalu and Barbados. Here, land will simply have to be abandoned and left to disappear beneath the sea, creating a growing number of **environmental refugees**. The major issue will be one of deciding where they should go to start a new life.

> **Environmental refugees:** People and communities forced to abandon their homes due to natural processes. These processes may be sudden, as with landslides or volcanic eruptions, or gradual, such as coastal erosion and rising sea levels.

Now test yourself

TESTED

29 Do you agree that the consequences of coastal change for communities are directly proportional to the density of population in the coastal zone?

Answer on p. 217

Different approaches to managing coastal risks

REVISED

Hard engineering

The traditional approach to dealing with the threat of coastal recession or flooding has been the hard-engineering approach – that is, to make extensive use of concrete, stone and steel to provide the required protection.

The main types of hard-engineering sea defence are:
- sea walls: made of reinforced concrete
- rip-rap (rock armour): huge rock boulders piled up at the base of a sea wall
- rock breakwaters: usually built of huge boulders, but offshore
- revetments: stone, timber or interlocking concrete structures on dune faces and mud banks
- groynes: vertical stone or timber fences built at 90 degrees to the coast and spaced along a beach.

Only the first of these, the sea wall, is widely used in the context of combatting coastal flooding. All the others are employed to deal with coastal erosion.

The hard-engineering approach has both advantages and disadvantages:
- It is obvious to those at risk that something is being done to protect them.
- It can be a one-off action that protects for decades.

However:
- Construction and on-going maintenance costs are high.
- Even very carefully designed engineering solutions can fail.
- The hard engineering is not visually attractive.
- Coastal ecosystems can be badly affected.
- Defences in one location can have adverse effects further along the coast in the direction of longshore drift.

Soft engineering

Soft engineering aims to work with natural processes to reduce the threats of coastal erosion and flooding. Because of this, it has two advantages over hard engineering: it is less visually intrusive and cheaper in the long term.

Examples of soft engineering include:
- beach nourishment: this is achieved by topping up beaches with sediment transported from elsewhere
- cliff stabilisation: stabilisation can be achieved by planting vegetation through a tough, flexible membrane that holds soil and often rock in place; it can also be achieved by regrading and reducing cliff slopes
- dune stabilisation: dunes are an effective form of coastal defence, but they are easily degraded; they can be stabilised by planting marram grass and by constructing relatively cheap dune fencing.

> **Exam tip**
>
> You need to learn the economic and environmental advantages and disadvantages of hard and soft engineering strategies on coasts. Sustainability is a vital criterion when judging both approaches.

> **Synoptic theme**
>
> Remember that any human intervention or action in the coastal system is likely to have an impact elsewhere. These impacts may be indirect and may have unforeseen consequences, such as accelerating erosion on other parts of the coast.

Sustainable coastal management

Today's threats to coastal areas of a rising sea level and an increasing frequency in storms call for a different and more comprehensive approach to the management of the coastal zone. That management needs to be sustainable in the sense of ensuring that the coastal zone and all its inhabitants have a reasonably secure future. Figure 2.19 gives a flavour of what is required. It also requires making full use of the concept of integrated coastal zone management (ICZM).

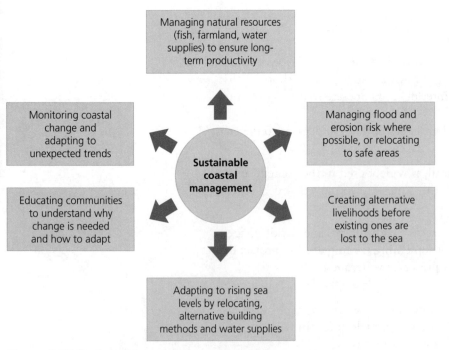

Figure 2.19 **Sustainable coastal management**

Exam practice answers and quick quizzes at **www.hoddereducation.co.uk/myrevisionnotes**

Integrated coastal zone management

REVISED

Key concept

ICZM dates from the Rio Earth Summit in 1992. The concept has three key features. It recognises:
- that the entire coastal zone needs to be managed, not just the zone where the breaking waves are causing erosion or flooding
- the importance of the coastal zone to people's livelihoods and well-being
- the need to make the management of the coast sustainable.

So ICZM is a joined-up or holistic approach to coastal management which must:
- plan for the long term
- involve all stakeholders and ensure they have a say in any policy decisions
- follow an 'adaptive' approach to unforeseen changes
- try to work with natural processes rather than against them.

ICZM works well with the concept of littoral cells (also referred to as sediment cells), which are natural subdivisions of the coastline containing sediment sources, transport paths and sinks. Each cell is isolated from adjacent cells and can be managed as a holistic unit. The coastline can be divided up into littoral cells and each managed as an integrated unit.

The coastline of England and Wales involves eleven cells. Each cell is managed by a shoreline management plan (SMP), which may be subdivided into sub-cells.

The hard part of **ICZM** is the decision-making process to decide what actions to take. In the UK, the critical decision concerns which one of four different management options to follow:
- No active intervention – the coast is left to erode or flood.
- Hold the line – build coastal defences so that the shoreline remains the same over time.
- Managed realignment – allow the coastline to change but in a controlled way (sometimes known as 'managed retreat').
- Advance the line – build new coastal defences on the seaward side of the existing coastlines. This usually involves land reclamation.

The choice of management option (the decision making) is generally not straightforward and depends on a number of factors:
- the economic value of the assets that might be protected
- the technical feasibility of different engineering solutions
- the environmental sensitivity
- the cultural and ecological value of the land that might be protected
- pressure from local communities, developers and environmental groups.

Any decision also needs to be informed by the results of objective investigations, such as cost–benefit analysis (CBA) and environmental impact assessments (EIAs).

It is clear that ICZM rarely pleases everyone. There are embedded conflicts of interest. Dissatisfaction will also be rooted in the fact that local councils and government have limited funds, so not all coastal areas can be protected. Coastal managers are forced to prioritise which areas should be protected and which will have to be ignored.

In many developed countries, there are good frameworks for effective coastal management. The same cannot be said for developing countries, despite the risks of coastal recession and flooding, and the huge populations, many large cities and great amounts of economic wealth concentrated in their coastal zones.

> **Exam tip**
>
> Coastal management is often controversial, with winners and losers. You should learn exemplars that will allow you to put forward a balanced and well-reasoned discussion.

Synoptic theme

It is unfortunate that not all the players in coastal environments share the same attitudes. What they seek ranges from the outright exploitation of the coast to strict conservation.

Revision activity

You are advised to make some notes on two coastal management studies, one in the UK and the other in a developing country (the Bangladeshi coastline is a good example to use). Your notes should focus on i) the nature of the coastal risk – recession or flooding, ii) identifying what is at risk – number of people, economic activities, etc., iii) management actions, if any, to date, and iv) identifying or anticipating the main winners and losers.

Now test yourself

TESTED ☐

30 What are sediment cells and why is it important that coastal management recognises the existence of such cells?

Answer on p. 217

Skills reminder

You should be familiar with the following skills and techniques used in the geographical investigation of coasts:
- interpreting GIS maps, satellite images and aerial photos
- field sketching
- field measurements – erosion, beach characteristics, coastal vegetation
- measures of central tendency
- student's *t*-test
- index of diversity
- cost–benefit analysis
- environmental impact assessment.

Exam practice answers and quick quizzes at **www.hoddereducation.co.uk/myrevisionnotes**

Exam practice

AS

1 (a) Name **one** type of mass movement that occurs on sea cliffs. (1)
  (b) Study Figure 1.

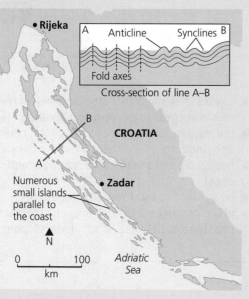

Figure 1 The coast of Croatia

  (i)   What is the name usually given to this type of coastline? (1)
  (ii) What is the approximate length of the transect between A and B? (1)
  (iii) Suggest how this coastline was formed. (3)
  (c) Explain the origins of sand dunes. (4)
  (d) Explain the causes and consequences of storm surges. (6)
  (e) Assess the value of a soft-engineering approach to the management of coastal erosion. (12)

A-level

2 (a) Study Figure 2.

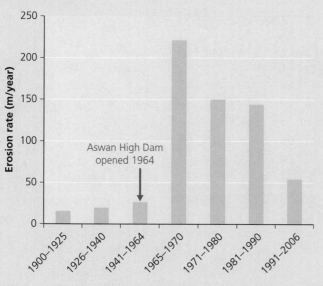

Figure 2 Erosion rates at Rosetta in the Nile Delta, Egypt, 1900–2006

(i) Explain the impact of the construction of the Aswan High Dam on the erosion rate at Rosetta in the Nile Delta. (6)

(ii) Explain the importance of coastal vegetation. (6)

(b) Explain how waves affect beach morphology. (8)

(c) Assess the risks and threats posed by the current rise in sea level. (20)

Answers and quick quiz 2B online

ONLINE

Summary

You should now have an understanding of:

- the physical and human importance of the coastal zone
- the physical variety of coastlines and the significance of geology
- the role of coastal vegetation
- the impacts of waves on the coast
- coastal erosion and its landforms
- coastal deposition and its landforms
- sub-aerial processes at work on cliffs

- coastal risks and threats – sea-level changes, storm surges, recession and flooding
- the human consequences of coastal recession and flooding
- hard, soft and sustainable management of the coast
- integrated coastal zone management
- place context studies of sample stretches of coastline in the UK and in the developing world.

3 Globalisation

Globalisation is the process in which places and people become more connected with each other than they used to be.

The causes and acceleration of globalisation

The acceleration of globalisation

> **Key concept**
>
> **Globalisation** is the latest chapter in a long story of how people and states have become more connected in four main ways:
> - Economic globalisation – through the growth of **transnational corporations (TNCs)** and information and communications technology (ICT).
> - Social globalisation – through international immigration, global improvements in education and health, and social interconnectivity.
> - Political globalisation – through the growth of trading blocs, free trade agreements and global organisations.
> - Cultural globalisation – through 'successful' Western cultural traits dominating in some territories, **glocalisation** and hybridisation, and the acceleration of circulation of ideas and information.

> **Transnational corporations (TNCs):** Businesses whose operations are spread across the world, operating in many nations as both makers and sellers of goods and services.
>
> **Glocalisation:** The changing of the design of products to meet local tastes or laws.

Global connections, flows and interdependence

Modern **globalisation** differs from that which preceded it by:
- a lengthening of connections between people and places, with products sourced from further away than ever before
- a deepening of connections, with the sense of being connected to other people and places now penetrating more deeply into almost every aspect of life
- faster connections, with people able to talk to one another in real time, using new technologies, or travelling quickly between continents using jet aircraft.

Globalisation involves building up networks of places and their populations. The connections between places represent different kinds of network flow. These flows are movements of:
- capital: at a global scale, major capital (money) flows are routed daily through the world's stock markets
- commodities: valuable raw materials such as fossil fuels, food and minerals have always been traded between nations
- information: the internet has brought real-time communication between distant places, allowing goods and services to be bought at the click of a button
- tourists: budget airlines have brought a 'pleasure periphery' of distant places within easy reach for the moneyed tourists of high-income nations

● migrants: the permanent movement of people still faces the greatest number of obstacles due to border controls and immigration laws.

Developments in transport and trade

In the nineteenth and twentieth centuries, developments in transport and trade went hand in hand (Figure 3.1). Improvements in transport allowed the volume and value of trade to increase and therefore some countries maintained their competitive edge through continued transport innovation.

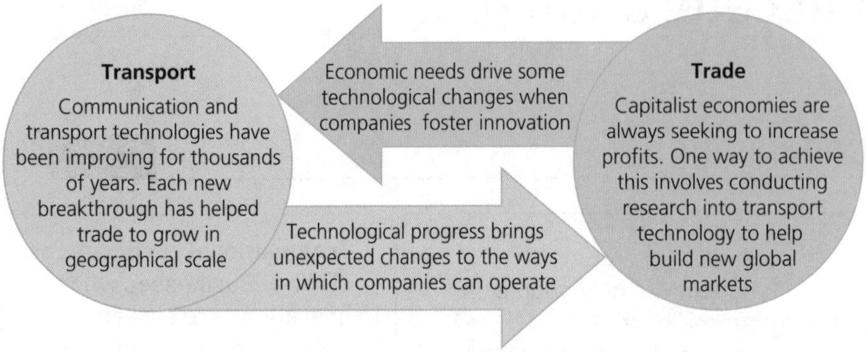

Transport
Communication and transport technologies have been improving for thousands of years. Each new breakthrough has helped trade to grow in geographical scale

Economic needs drive some technological changes when companies foster innovation

Technological progress brings unexpected changes to the ways in which companies can operate

Trade
Capitalist economies are always seeking to increase profits. One way to achieve this involves conducting research into transport technology to help build new global markets

Figure 3.1 The interrelationship between trade and transport growth

Important innovations in transport included:
● steam power: steam ships (and trains) moved goods and armies quickly along trade routes into Asia and Africa
● railways: in the 1800s, railway networks expanded globally; today, railway building remains a priority for governments across the world
● jet aircraft: the arrival of the intercontinental jet aircraft in the 1960s made international travel more commonplace
● container shipping: the 'backbone' of the global economy since the 1950s, with around 200 million individual container movements taking place each year.

These innovations have resulted in a shrinking world, with distant places feeling closer and taking less time to reach.

ICT and mobile phone use in the twenty-first century

Technology is used by different players in a vast array of ways that contribute to globalisation. Some of these include:
● telephone and the telegraph: core technologies for communicating across distance
● broadband and fibre optics: enormous flows of data are conveyed across the ocean floor through fibre optic cables
● GIS and GPS: satellites continuously broadcast position and time data to users throughout the world
● the internet, social networks and Skype: connect people and places across the globe.

> **Exam tip**
>
> Be sure you know the four ways in which people and places have become more connected.

Now test yourself

TESTED

1 How have developments in transport and trade contributed to a shrinking world?

Answer on p. 217

Exam practice answers and quick quizzes at **www.hoddereducation.co.uk/myrevisionnotes**

The politics and economics of globalisation

The work of international organisations

Synoptic theme

Three intergovernmental organisations (IGOs) are major players in the promotion of the global economy (Table 3.1):

- the International Money Fund (IMF)
- the World Bank
- the World Trade Organization (WTO).

Other important players are TNCs and national governments.

Table 3.1 Evaluating the role of international organisations in globalisation

	IMF	World Bank	WTO
Role in globalisation	Based in Washington, DC, the IMF channels loans from rich nations to countries that apply for help. In return, the recipients must agree to run free market economies that are open to outside investment. As a result, TNCs can enter these countries more easily. The USA exerts significant influence over IMF policy despite the fact that it has always had a European president.	The World Bank lends money on a global scale and is also headquartered in Washington, DC. In 2014, a US$470 million loan was granted to the Philippines for a poverty-reduction programme, for instance. The World Bank also gives direct grants to developing countries – in 2014, help was given to Democratic Republic of the Congo to kick-start a stalled mega-dam project.	The WTO took over from the General Agreement on Tariffs and Trade in 1995. Based in Switzerland, the WTO advocates trade liberalisation, especially for manufactured goods, and asks countries to abandon protectionist attitudes in favour of untaxed trade (China was persuaded to lift export restrictions on 'rare earth' minerals in 2014).
Evaluation	IMF rules and regulations can be controversial, especially the strict financial conditions imposed on borrowing governments, which may be required to cut back on healthcare, education, sanitation and housing programmes.	In total, the World Bank distributed US$65 billion in loans and grants in 2014. However, like the IMF, the World Bank imposes strict conditions on its loans and grants. Controversially, all World Bank presidents have been US citizens.	The WTO has failed to stop the USA and the EU, for example, from subsidising their own food producers. This protectionism is harmful to farmers in developing countries which want to trade on a level playing field.

Now test yourself

2 How have IGOs contributed to globalisation?

Answer on p. 217

The attitudes and actions of national governments

National governments are also players in globalisation and help promote the growth of the global economy by:

- promoting **trade blocs** – for example, the European Union (EU) and the Association of Southeast Asian Nations (ASEAN)
- free-market liberalisation – for example, lifting restrictions on the way companies and banks operate

Trade blocs: Voluntary international agreements that encourage the free flow of goods and capital between member countries.

- encouraging business start-ups and allowing TNCs to grow in size and influence
- privatisation – allowing companies to take over important national services such as railways and energy supply.

The spread of globalisation into new global regions

Special economic zones (SEZs), government subsidies and changing attitudes to **foreign direct investment (FDI)** have played important roles in the acceleration of globalisation. This is well illustrated by the case of China. Its 'open door policy' has allowed urbanisation to fuel the growth of low-wage factories. These, in turn, have enabled China to become the 'workshop of the world'. The world's largest TNCs were quick to establish branch plants or trade relationships with Chinese-owned factories, in newly established coastal SEZs. Today, China is the world's largest economy and is evidence that global-scale free trade can sometimes cure poverty.

The effects of globalisation

REVISED

Uneven globalisation

Globalisation has affected some places more than others. The unevenness is due to variations in the policies and attitudes of governments – not all governments favour involvement in the global economy – physical factors, such as the availability of resources and accessibility, and TNCs' assessments of where the best business opportunities are.

Uneven levels of globalisation can be assessed using various measures and indicators. The Swiss Institute for Business Cycle Research (KOF) produces an annual 'index of globalisation' using a range of diverse data such as participation in UN peace-keeping missions and TV ownership (Figure 3.2). The A.T. Kearney World Cities Index ranks a city's 'business activity', 'cultural experience' and 'political engagement' to determine how well it is 'globally connected'.

Special economic zone (SEZ): An industrial area, often near a coastline, where favourable conditions are created to attract foreign TNCs.

Foreign direct investment (FDI): A financial injection made by a business into another country's economy, either to build new facilities (factories or shops) or to acquire, or merge with, a firm already based there.

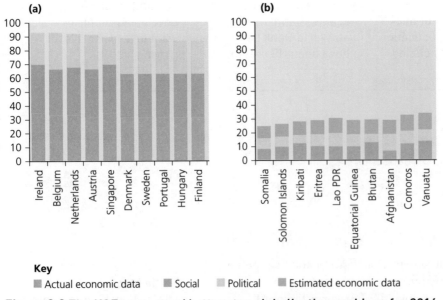

Key
■ Actual economic data   ■ Social   ■ Political   ■ Estimated economic data

Figure 3.2 The KOF top ten and bottom ten globalisation rankings for 2014

TNCs and globalisation

TNCs attempt to build their global businesses through the processes of:

- offshoring: moving parts of their production process (factories or offices) to other countries to reduce labour or other costs
- outsourcing: contracting another company to produce the goods and services they need rather than do it themselves
- global production networks: setting up chains of connected suppliers of parts and materials that contribute to the manufacturing or assembly of the consumer goods.

However, some parts of the world have benefited far more than others from FDI by TNCs because not all places are suitable sites of production for goods, for a range of physical and human reasons, and not all places have enough market potential to attract large retailers.

In order to maximise profit, many TNCs have, additionally, adapted their products to suit local tastes through the process of **glocalisation** in which products are adapted to suit:

- people's tastes
- religion and culture
- laws
- local interests
- lack of availability of raw materials.

> **Typical mistake**
>
> TNCs might look the same the world over, but on closer investigation many of the products they sell will have been adapted for local markets.

Now test yourself

TESTED

3 List the ways in which TNCs are important players in globalisation.

Answer on p. 217

A few of the very poorest nations of the world remain relatively switched off from global networks due to a range of physical, political, economic or environmental reasons. A prime example is North Korea; other examples include countries in the Sahel.

Now test yourself

TESTED

4 Why are some places 'switched off' from globalisation?

Answer on p. 217

> **Exam tip**
>
> Make sure that you can suggest a range of reasons why globalisation has affected some places and organisations more than others and that you can support your ideas with real place examples.

> **Revision activity**
>
> Make sure that you understand why globalisation has accelerated in recent decades.

Geographical impacts of globalisation

Winners and losers

REVISED

Globalisation has had a range of impacts on different countries, as well as on different groups of people and cultures through:

- **global shifts** in economic activity that have brought a range of environmental, economic and social impacts
- the increasing scale and pace of economic migration, resulting in a range of impacts to places at varying scales
- global and local cultural changes associated with globalisation.

> **Global shift:** The international relocation of different types of industrial activity, especially manufacturing industries.

Costs and benefits for emerging Asia

Across Asia, urban environments have been transformed by rapid industrialisation and the establishment of SEZs. Major economic, social and environmental changes associated with globalisation include:

- poverty reduction and waged work
- education and training
- undue pressure on resources and the environment
- growth of the built environment and unplanned settlements.

Environmental challenges in developing countries

Global shift has been driven by TNCs seeking low-cost locations for their manufacturing and refining operations. Lax regulation of pollution, environmental protection, and health and safety has been an attractive location factor for them.

Problems of deindustrialisation

Global shift creates challenges for developed countries too. **Deindustrialisation** in urban areas has fuelled a spiral of decline (Figure 3.3) involving:

- high unemployment due to the disappearance of 'old' manufacturing jobs
- rising crime that has now become the basis of an informal economy in poor urban areas
- depopulation as people migrate out of failing urban areas in large numbers, with those who stay often becoming trapped
- dereliction resulting from abandoned factories, closure of services, rundown housing and environmental neglect.

> **Revision activity**
>
> Make a table summarising the benefits and costs of globalisation in Asia.

> **Deindustrialisation:** The decline of regionally important manufacturing industries in terms of either workforce numbers or output and production measures.

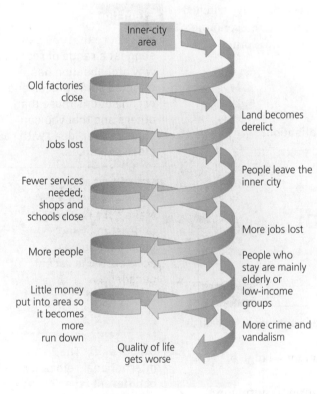

Figure 3.3 The inner-city spiral of decline and deprivation

> **Exam tip**
>
> You must be able to evaluate a range of costs and benefits created by globalisation.

> **Exam tip**
>
> Make sure that you understand that global shift has created challenges for some people living in developed countries.

Now test yourself

TESTED

5 How has global shift created winners and losers?

Answer on p. 217

The increasing scale and pace of economic migration

REVISED

Rural–urban migration and megacity growth

A megacity is home to 10 million people or more. In 1970 there were just three; by 2025 there will be 30 (Figure 3.4). They grow through a combination of rural–urban migration and natural increase. The causes of urban–rural migration are:

- urban pull factors: the main factor almost everywhere is employment. Urban areas offer the hope of promotion and advancement into occupational roles that are non-existent in rural areas. Schooling and healthcare may be better in urban areas, making cities a good place for young migrants with aspirations
- rural push factors: the main factor is usually poverty, aggravated by population growth (not enough jobs for those who need them) and land reforms (unable to prove they own their land, subsistence farmers must often relocate to make room for TNCs and cash crops)
- 'shrinking world' technology: rural dwellers are gaining knowledge of the outside world and its opportunities and transport improvements have removed intervening obstacles to migration.

> **Exam tip**
>
> Make sure you have some details about a specific megacity. Mumbai would be a good one.

> **Revision activity**
>
> Make brief notes about the distribution of megacities shown in Figure 3.4.

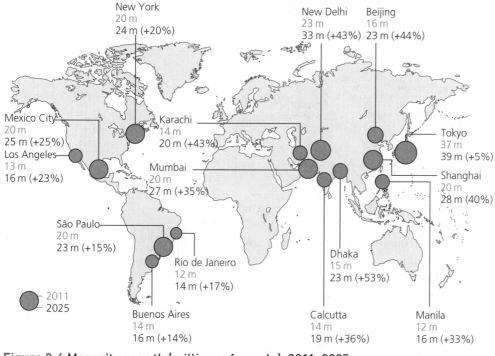

Figure 3.4 Megacity growth (millions of people), 2011–2025

Rapid urban growth, particularly in megacities, is creating a range of social and environmental costs:

- inadequate provision of housing
- limited access to education and healthcare

- pollution of water and air
- loss of farmland.

Now test yourself

TESTED

6 How has rural–urban migration contributed to the growth of megacities?

Answer on p. 217

International migration into global hubs

> **Key concept**
>
> A **global hub** is a highly globally connected city, or the home region of a large, globally connected community that has become a focal point for activities with a global influence, such as trade, business, international governance, education and research. Unlike a megacity, a global hub is recognised by its influence rather than its population size (Figure 3.5). Flows of money, goods and workers help link the world's global hubs to form a network of important places.

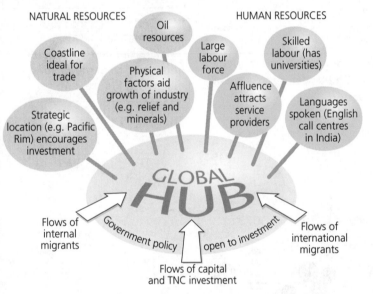

Figure 3.5 Resources involved in the growth of global hubs

Two types of international migration have contributed to the growth of **global hubs**:

- Elite international migration involving highly skilled and/or socially influential individuals (Figure 3.6). Some elite migrants live as 'global citizens' and have multiple homes in different countries. They encounter few obstacles when moving between countries. Example: migration of Russian oligarchs to London.
- Low-waged international migration in which people are drawn towards global hubs in large numbers of both legal and illegal immigrants working for low pay in kitchens, on construction sites or as domestic cleaners. Example: migration of cheap labour from India to the UAE, Saudi Arabia and the Philippines.

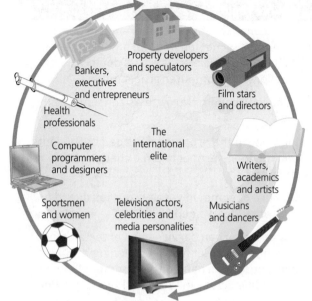

Figure 3.6 Elite migration

Now test yourself   TESTED ☐

7 Explain how international migration has increased in global hub cities.

Answer on p. 217

The costs and benefits of migration

> **Key concept**
>
> **Interdependence** is very much one of the products of globalisation. It builds up as countries become increasingly reliant on each other in terms of trade as well as financial and political connections. While interdependence offers economic and political security, it does have a downside, namely that a country's continued well-being depends on the economic health of those involved in the same network of interdependence.

International migration increases countries' **interdependence**. It also creates a wide range of costs and benefits, which are summarised in Table 3.2.

Table 3.2 Selected costs and benefits of migration for source and host countries

	Host region	Source region
Benefits	Fills particular skills shortages (e.g. Indian doctors arriving in the UK in the 1950s). Economic migrants willingly do labouring work that locals may be reluctant to do (e.g. Polish workers on farms around Peterborough). Working migrants spend their wages on rent, benefiting landlords, and pay tax on legal earnings. Some migrants are ambitious entrepreneurs who establish new businesses employing others (in 2013, 14 per cent of UK business start-ups were migrant-owned).	Migrant remittances can contribute to national earnings significantly (in 2014, remittances made up 25 per cent of Nepal's national earnings). There will be less public spending on housing and health (in 2004, prior to joining the EU, unemployment in Poland was 20 per cent; it has since halved). Migrants or their children may return, bringing new skills (young British Asians have relocated to India to start health clubs and restaurant chains). Some government spending costs (education, health) are transferred to the host region.

	Host region	Source region
Costs	Social tensions arise if citizens of the host country believe migration has led to a lack of jobs or affordable housing (a view adopted by the UKIP party and some UK newspapers).	The economic loss of a generation of human resources, schooled at government expense, including key workers such as doctors, teachers and computer programmers.
	Political parties change their policies to address public concerns (e.g. pledges to reduce migration).	Reduced economic growth as consumption falls.
	There may be local shortages of primary school places due to a natural increase among a youthful migrant community (e.g. London boroughs that have become Eastern European migration 'hotspots').	Increase in the proportion of aged dependants and the long-term economic challenge this creates.
		Closure of some university courses due to a lack of students aged 18–21.
	New markets can develop for ethnic food (e.g. Korean food markets in Los Angeles), bringing visible changes to the urban built environment.	Closure of urban services and entertainment with a young adult market, bringing decline and dereliction to urban built environments (many nightclubs closed in Warsaw, Poland, in 2004).

Revision activity

Note the costs and benefits of migration for both source and host countries.

Now test yourself

TESTED

8 What is interdependence, and what are its advantages?

Answer on p. 217

The emergence of a global culture

REVISED

Key concept

One of the products of globalisation is the **diffusion of a global culture**. It is being spread by the mass media, the internet and migration. As it spreads, it tends to erode traditional cultures, but occasionally cherry picks, adapts and absorbs selected parts of those cultures. Diagnostic features of the emerging global culture are:
- emphasis on consumerism
- belief in capitalism and personal wealth
- Anglo-Saxon and white
- English (American) the dominant language
- 'American' in terms of tastes and traits – diet, dress, entertainment, etc.

Cultures change and evolve over time naturally, but globalisation has accelerated the rate of change for many places. It has led to the evolution and **diffusion of a global culture**. Perspectives differ on both the degree of cultural change that is occurring and its desirability.

Exam practice answers and quick quizzes at **www.hoddereducation.co.uk/myrevisionnotes**

Cultural diffusion and its causes

Much cultural change has been the product of gradual **cultural diffusion**. Today, however, it is being precipitated, even forced, by **cultural imperialism**. Countries like the USA play a role in bringing cultural change to other places through their use of soft power and large influence over global media and entertainment.

> **Cultural diffusion:** The gradual spread of culture from an influential civilisation.
>
> **Cultural imperialism:** The practice of promoting the culture/language of one nation in another. It is usually the case that the former is a large, economically or militarily powerful nation and the latter is a smaller, less affluent one.

> **Synoptic theme**
>
> The TNCs and global media companies are foremost among the players promoting a global culture. Tourists and migrants can also play a part.

Table 3.3 evaluates the influence of the three different players in the diffusion of a global culture.

Table 3.3 Evaluating the emergence of a global culture

Factor	TNCs	Global media	Migration and tourism
Influence	The global dispersal of food, clothes and other goods by TNCs has played a major role in shaping a common culture. Some corporations, such as Nike, Apple and Lego, have 'rolled out' uniform products globally, bringing cultural change to places.	Media giant Disney has exported its stories of superheroes and princesses everywhere. Western festivals of Halloween and Christmas feature prominently in its films. The BBC helps maintain the UK's cultural influence overseas (especially the World Service radio station).	Migration brings enormous cultural changes to places. Europeans travelled widely around the world during the age of empires, taking their languages and customs with them. Today, tourists introduce cultural change to the distant places they visit.
Evaluation	When TNCs engage with new markets and cultures, they often adapt their products and services to suit different places better. As a result, the products that are sold in different places increasingly reflect local cultures. You will be familiar with examples of this, such as McDonald's menus. In your view, is glocalisation merely a sophisticated form of cultural imperialism?	Other places gain a 'window' on American and British culture through shows such as the period drama 'Downton Abbey'. However, many reality and celebrity shows, such as 'Strictly Come Dancing', are entirely refilmed for different national markets. Also, there are many non-Western influences on global culture, including the TV channels Russia Today and Qatar's Al Jazeera. Japanese children's TV has been highly influential, notably through Pokémon.	Migrants can affect the culture of host regions, but the change may be only partial. British migrants took their language and love of cricket to many places but often had little effect on other cultural traits, notably religion. When carrying out an evaluation, it is important to ask whether cultural changes for places are superficial or more meaningful. We can also explore the effect of out-migration on the culture of source regions.

> **Revision activity**
>
> Summarise some of the impacts of cultural diffusion and how this has helped create and spread an increasingly 'Westernised' global culture.

The costs of cultural erosion

The idea that a largely Westernised global culture is emerging as a result of cultural erosion in different places is called **hyper-globalisation**.

This is often seen as a negative development by some people who are concerned that languages around the world are disappearing as use of English continues to spread. They also fear a global trend in the devaluing of ecosystems, at varying scales. The capitalist philosophy behind globalisation makes economic growth its primary goal and economies which do not grow are often deemed failures.

The opposite view is that globalisation and cultural erosion can bring positive change on a worldwide scale, with the emergence of a global culture that values equality, freedom of expression and reduced discrimination on the grounds of gender, sexuality or disability.

The loss of tribal lifestyles and rights in Papua New Guinea and Amazonia are prime examples of cultural erosion.

Concern about the impacts of globalisation

There is growing opposition to globalisation on a range of geographical scales. Individuals, pressure groups and governments may all have some degree of concern about the cultural impacts of globalisation. It is important to realise that culture can profoundly affect attitudes to all sorts of other issues, such as resources, the environment and heritage.

Typical mistake

Cultural erosion does not always have to be negative. Some people will readily embrace the cultural changes that globalisation can bring.

Revision activity

It is important that you have some notes about the impacts of cultural erosion (loss of language, traditional food, music, clothes, social relations).

Synoptic theme

There are strongly polarised views (attitudes) on globalisation, particularly its environmental and cultural impacts. This is leading to the establishment of protest movements – both pro- and anti-globalisation. Strong attitudes against aspects of globalisation exist in various environmental and cultural organisations, such as Greenpeace and UNESCO.

Exam tip

Globalisation is often cited as 'being to blame' for a range of problems – but there are many benefits to be had as well.

Now test yourself   TESTED

9 Suggest reasons why there is growing concern about the impacts of globalisation.

Answer on p. 217

The challenges of globalisation

Globalisation, development and the environment   REVISED

The **development gap** (the range of values between the world's very richest and poorest people and countries) has increased as globalisation has accelerated. Overall, the global economy has grown enormously – far faster than its population.

The already-rich take a disproportionately large share of each year's new economic growth. In contrast, the wealth and incomes of the majority of people have grown more slowly over time. Globalisation has not pushed large numbers of people into absolute poverty; instead, global poverty has been halved since the introduction of the Millennium Development Goals in 2000.

Economic and social development measures

Development is measured in many different ways using both single and composite measures:

- Gross national income (GNI) per capita is the mean average income of a population. It is calculated by taking an aggregate source of income for a country and dividing it by population size to give a crude average.
- Gross domestic product (GDP) is a widely used aggregate measure. It is the final value of the output of goods and services inside a nation's borders, and is most telling if expressed in per capita terms.
- Economic sector balance – a country's or region's economy can be crudely divided into four economic sectors whose relative importance changes as a country develops. The balance shifts from the primary to the tertiary sector.
- The human development index (HDI) is a composite measure that ranks countries according to economic criteria (GDP per capita, adjusted for purchasing power parity) and social criteria (life expectancy and literacy).
- The gender inequality index (GII) is another composite index devised by the UNDP. It measures gender inequalities related to women's reproductive health, empowerment and labour force participation.
- Environmental quality is often poor in developing and emerging economies. It usually improves as economic and social development occurs and places make the transition from industrial to post-industrial forms of economic activity. Possible measures here include levels of air pollution.

Now test yourself

TESTED

10 What are the benefits of assessing development by social measures?

Answer on p. 217

Widening inequalities

The steep rise in world money supply during the era of globalisation has been accompanied by a changing spatial pattern of global wealth:

- Average incomes have risen on all continents since 1950, but only very slowly in the poorest parts of Africa.
- The great gains made by European and North American nations over the same time period have resulted in a widening of the average income gap between people living in the world's wealthiest and poorest countries.
- Absolute poverty has fallen worldwide.
- Many countries have advanced from low-income to middle-income status since the 1970s.
- There is a growing wealth divide within some nations.

> **Key concept**
>
> The **Gini coefficient** is a measure of income inequality given as a
> number between zero and 100. The higher the value, the greater the
> degree of income inequality. A value of zero suggests that everyone
> has the same income whereas a value of 100 would mean a single
> individual receives all of a country's income.

Analyses using the **Gini coefficient** show only a rough correlation
between a country's level of development and the degree of income
equality. The coefficient can also be used to assess another important
dimension, namely inequalities within individual countries.

Major environmental issues linked with globalisation include climate
change and biodiversity loss:
● The transformation of 40 per cent of the Earth's terrestrial surface into
  productive agricultural land has led to habitat loss and biodiversity
  decline on a continental scale.
● The negative impacts of large agribusiness operations penetrate deeply
  into many of the world's poorer regions.
● Intensive cash cropping, cattle ranching and aquaculture bring
  damaging environmental effects ranging from groundwater depletion
  to the removal of forests.

Contrasting trends in economic development and environmental management

Figure 3.7 shows that since 1820 all global regions have experienced a
rise in per capita GDP. But of much greater significance is that the rise in
Africa, Asia and Latin America has been substantially less compared with
that in Western Europe and the USA. Even since 1970, those two regions
have pulled away from the rest of the world. The beginning of Asia's
take-off can be seen since 2000.

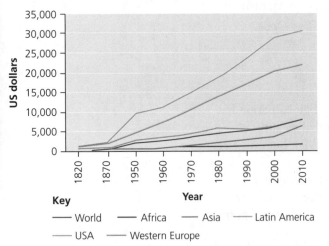

Figure 3.7 Per capita GDP growth for global regions since 1820

The environmental issues indicated earlier have been most keenly felt in
developing and emerging countries. These have been the environmental
losers. In developed countries, there is now a growing awareness of the
environmental risks associated with globalisation, and attempts are being
made through environmental management to curb such risks.

Now test yourself

TESTED

11 Why have some regions benefited more from globalisation than others?

Answer on p. 217

Social, political and environmental tensions of globalisation

REVISED

The speed and scale of globalisation have become such as to generate a range of tensions – social, political and environmental.

Cultural mixing

Large volumes of international migration are a feature of globalisation. They have been encouraged by open borders, deregulation and FDI. Together with the promotion of a global culture, they have created culturally mixed societies. They have also given rise to thriving **diasporas** throughout the world.

> **Diaspora:** The dispersion or spread of a group of people from their original homeland.

Tensions can easily develop between immigrant groups and host communities, particularly if there are differences in ethnicity. Tensions fester on fears that the immigrants are taking over jobs and homes and overloading social services. Immigrants feel that they are being discriminated against and this also adds to the tension. Calls are made to control immigration flows.

> **Exam tip**
>
> Migration can be a rather sensitive issue as people have different opinions. Be careful to balance different attitudes from a geographical viewpoint.

In some EU states, nationalist parties have been gaining support because they often oppose immigration; some reject multiculturalism and openly embrace fascism.

Now test yourself

TESTED

12 How has globalisation contributed to the growth of culturally mixed societies?

Answer on p. 217–8

Controlling the spread of globalisation

Not all governments greet globalisation with open arms. Some have tried to prevent or control the global flows of people, goods and information by:

- strengthening laws to limit numbers of **economic migrants**. However, illegal immigration can be hard to tackle. Since 1948, the Universal Declaration of Human Rights (UDHR) has guaranteed refugees the right to seek and enjoy asylum from persecution.
- limiting citizens' freedom to access online information. However, a 'dark web' exists and is harder to control.
- trade protectionism, which is still common – the illegal smuggling of both legal and illegal commodities can be very difficult to control.

> **Economic migrant:** A person who moves from one country to another in order to find work and improve their standard of living and quality of life.

> **Typical mistake**
>
> Despite some internet restrictions, many people in China enjoy a wide range of freedoms and some argue that control of the internet is good governance.

Internet censorship still exists in China and this is part of the government's strategy to stifle criticism and censorship. Although Facebook, Twitter and YouTube remain unavailable there due to the 'great firewall of China', more than 400 million Chinese citizens interact with each other using local social media sites, such as Youku Tudou.

Now test yourself

TESTED

13 List the ways in which some countries have attempted to control the spread of globalisation.

Answer on p. 218

Resource nationalism and protecting cultures

Some countries have adopted **resource nationalism** as a response to globalisation. For example:

- the Venezuelan government took control of ExxonMobil and ConocoPhillips oil operations in their country
- the Canada-based First Quantum was forced to hand over 65 per cent ownership of a US$550 million copper mining project in Democratic Republic of the Congo to the country's government
- until recently, resource nationalism in China took the form of restrictions on rare earth exports.

Particular cultural groups within a nation, for example the six indigenous peoples in Canada, may sometimes take a view on whether global forces should be allowed to exploit their resources. Opposition can be strong when an important landscape is threatened by resource extraction.

> **Resource nationalism:**
> A growing tendency for state governments to take measures ensuring that domestic industries and consumers have priority access to the national resources found within their borders.

> **Revision activity**
>
> Draw up a table to summarise some of the social, political and environmental tensions that have resulted from globalisation.

Globalisation, sustainability and localism

REVISED

Globalisation is associated with a range of rising environmental stresses related to population pressure on food, water and energy.

The 'local sourcing' solution

Some environmentally minded citizens adopt an **ethical consumption** strategy by purchasing locally sourced food and commodities. They boycott supermarket products with high food miles. Local pressure groups play an important role in promoting local sourcing (Table 3.4). One of the greatest benefits of local sourcing of food is the reduction in food miles.

> **Ethical consumption:**
> This occurs when the consumer takes into account the costs (economic, social and environmental) of producing food and goods, and of providing services.

Table 3.4 The costs and benefits of local sourcing

	Costs	Benefits
Consumers	Local sourcing of everyday meat and vegetables can be very expensive, especially for people on low incomes.	Many small producers in the UK have adopted organic farming methods. Crops are grown using fewer pesticides, which could have health benefits.
Producers	Less demand from UK consumers for food from producer countries means arrested economic development for places such as Ivory Coast.	UK farmers have moved up the value chain by manufacturing locally sourced items, including jams, fruit juices and wine.
Environment	Tomatoes in the UK are grown in heated greenhouses and polytunnels during winter, resulting in a larger carbon footprint than imported Spanish tomatoes.	The 1992 Rio Earth Summit introduced the slogan 'Think global, act local'. Local sourcing sometimes helps people reduce their carbon footprint.

Synoptic theme

Local pressure groups can do much through a range of actions to promote localism and ethical consumption.

Now test yourself

TESTED

14 How can local sourcing reduce some of the negative environmental impacts associated with globalisation?

Answer on p. 218

Ethical consumption and fair trade

While consumers benefit economically from the global shift towards cheaper goods, many have ethical concerns about the social costs of worker exploitation. 'Opting out' of buying globally sourced commodities is very hard to do in practice. However, ethical purchases are increasingly available thanks to the work of NGOs, charities and a growing number of businesses with a 'social responsibility' agenda, as can be seen in Table 3.5.

Table 3.5 Evaluating ethical consumption schemes

	Actions	Evaluation
Fair trade	The Fairtrade Foundation's certification scheme offers a guaranteed higher income to farmers and some manufacturers, even if the market price changes.	Fair trade goods let shoppers know that what they spend will find its way into the pay packets of poor workers – although not all shoppers will pay more.
	Examples of fair trade produce include coffee, chocolate, bananas, wine and even clothing items such as jeans.	However, as the number of schemes grows, it becomes harder to ensure that money has been correctly distributed.
	The Waitrose Foundation has also embraced fairer trading principles by improving pay for farmers in its supply chains.	It is not possible for all the world's farmers to join a scheme offering a high fixed price for potentially unlimited crop yields.

	Actions	Evaluation
Supply chain monitoring	Large businesses increasingly accept the need for corporate social responsibility. The largest TNCs have thousands of suppliers; this increases the risk of branded products being linked with worker exploitation. Apple investigated its iPhone touchscreen supplier, Lianjian Technology, whose workers were poisoned by a chemical cleaning agent.	Firms such as Gap and Nike now prohibit worker exploitation in their foreign factories, but it is difficult to monitor the working conditions and pay for the workforce of every single supplier they buy from. It is especially hard to control what happens in the workplaces of their suppliers.
NGO action	Charity War on Want helped South African fruit pickers. It flew a woman called Gertruida to a Tesco shareholder meeting in London. Gertruida explained that there was no toilet for female workers at the farm where she worked. Tesco told the farm it would use a different fruit supplier unless conditions improved.	NGOs have limited financial resources. This can inhibit the scale of what they can achieve, or result in slow progress. Although NGOs such as Amnesty International work hard to raise awareness of ethical issues, many people remain unaware of, or unconcerned about, worker exploitation.

Recycling and resource consumption

At the end of their useful life, manufactured goods are often sent as waste to landfill. An alternative is to recycle them. This reduces the rate at which new natural resources are used. The recycling process does itself, however, require the use of energy and water.

Recycling can be viewed as the first step towards the more ambitious goal of a circular economy (Figure 3.8) as well as reducing the ecological footprint.

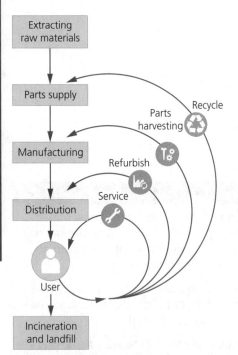

Figure 3.8 The 'circular' economy

> ### Synoptic theme
> There is uncertainty about the exact environmental consequences of different patterns of consumption.

> ### Revision activity
> Summarise how ethical and environmental concerns about unsustainability have led to increased localism and awareness of the impacts of a consumer society.

Now test yourself

15 What are the benefits of recycling? Is there a downside?

Answer on p. 218

Skills reminder

You should be familiar with the following skills and techniques used in the geographical investigation of globalisation:
- using proportional flow lines to show networks of flows
- ranking and scaling data to create indices
- analysing human and physical features on maps to understand a lack of connectedness
- using population, deprivation and land-use data sets to quantify the impacts of deindustrialisation
- using proportional flow arrows to show global movement of migrants from source to host areas
- analysing global TNC and brand value data sets to quantify the influence of Western brands
- critically using World Bank and United Nations (UN) data sets to analyse trends in human and economic development, including the use of line graphs, bar charts and trend lines
- plotting Lorenz curves and calculating the Gini coefficient.

Exam practice

AS

1 (a) What is the term used when a TNC contracts another company to produce the goods or service that it needs? (1)

  (b) Study Figure 1.

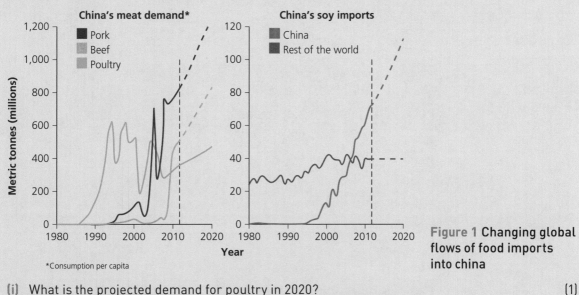

Figure 1 **Changing global flows of food imports into china**

  (i)   What is the projected demand for poultry in 2020? (1)
  (ii) Calculate how much more pork than beef will be in demand in 2020. (1)
  (iii) Suggest **one** reason for why meat demand is increasing in China. (3)
  (c) Explain how transport and trade have contributed to globalisation. (4)
  (d) Explain how global shift has created winners and losers for people. (6)
  (e) Assess the extent to which concerns about unsustainability have increased localism and awareness of the impacts of a consumer society. (12)

A-level

2 (a) Explain how political and economic decision making contribute to the acceleration
of globalisation. (4)
(b) Assess the extent to which social, political and environmental tensions have resulted
from the rapidity of global change caused by globalisation. (12)

Answers and quick quiz 3 online

ONLINE

Summary

You should now have an understanding of:
- globalisation, a long-standing process, which has accelerated because of rapid developments in transport, communications and businesses
- political and economic decision making – important factors in the acceleration of globalisation
- how globalisation has affected some places and organisations more than others
- the way in which the global shift has created winners and losers for people and the physical environment
- the scale and pace of economic migration, which have increased as the world has become more interconnected, creating consequences for people and the physical environment

- the emergence of a global culture, based on Western ideas, consumption and attitudes towards the physical environment – an outcome of globalisation
- how globalisation has led to dramatic increases in development for some countries, but also widening development gap extremities and disparities in environmental quality
- social, political and environmental tensions which have resulted from the rapidity of global change caused by globalisation
- ethical and environmental concerns about unsustainability which have led to increased localism and awareness of the impacts of a consumer society.

---

4 Shaping places

Option A Regenerating places

This option topic focuses on the economic and social changes affecting places, particularly within the UK, and the need for regeneration that often arises from these changes. Regeneration is aimed at reviving the flagging economies of places and improving the quality of life of local people. Two key concepts running through the whole of this topic are place and regeneration.

Key concepts

Place is a part of geographical space with a distinctive identity and character felt deeply by local inhabitants. The sense of each and every place derives from a unique mix of external connections, natural and human features in the landscape, and the people who happen to occupy it.

Regeneration involves positively transforming the economy of a place that has displayed symptoms of decline, making it viable and sustainable. It frequently goes hand in glove with rebranding (changing people's perceptions of a place in order to promote it) and reimagining (positively changing the standing and reputation of a place through specific improvements).

Now test yourself

TESTED

1 What is regeneration and why is it such an important issue in the UK?

Answer on p. 218

How and why places vary

Places are shaped by internal **connections** (between people, employment, services and housing) and external connections (such as government policies and globalisation). More specifically, place identity is strongly influenced by what places and their residents do for a living. The economic function of a place will also affect the type of work on offer and that work, in turn, will affect the type of employee. Compare, for example, the labour needs of a coastal resort, such as Torquay, with those of an industrial town, such as Sunderland.

> **Connection:** Any type of physical, social or online linkage between places. Places may keep some of their characteristics or change them as a result.

Economies vary from place to place

REVISED

Classifying economies and workers

A key factor in the creation and survival of places is their economy. This will affect important aspects such as identity, income and lifestyles, as well as the socio-economic composition of the local population.

There are four key employment sectors: primary, secondary, tertiary and quaternary industries. The emergence of the last of these has been relatively recent (say, since 1975). The UK has seen employment declines in the primary industries and secondary sectors linked to **deindustrialisation** and the global shift of manufacturing. The tertiary sector now totally dominates the economy. These sectoral shifts are mirrored in the economies of most places.

> **Deindustrialisation:** The process of economic and social change due to a reduction in the industrial capacity or activity of a country, region or city. A process widely experienced in the developed world with the global shift of manufacturing to emerging economies.

Now test yourself

TESTED ☐

2 How has sectoral employment in the UK changed over the past 150 years?

Answer on p. 218

The type of employment found in all sectors can be classified in several different ways:
- part–time/full–time
- temporary/permanent
- employed/self–employed.

The workers themselves can also be classified:
- employees with contracts (permanent or fixed)
- agency staff and volunteers
- self–employed (freelancers, consultants and contractors)
- skilled/semi–skilled/unskilled.

> **Exam tip**
>
> Make sure you know the different employment sectors and the types of worker, and how the sectoral balance of the UK's economy has changed.

Social impacts

These distinctions on the basis of type of employment and worker have profound impacts on people's lives, particularly on:
- health: for example, long working hours in manual jobs, such as construction and agriculture, lead to an increased risk of accidents and poor health
- life expectancy: there is a whole range of work-related factors here, such as levels of stress, exposure to risk, diet, etc.
- education: for example, children from poor working-class families still tend to underachieve and are often denied the chance of going to university
- lifestyles: higher wage and salary levels mean more disposable income to be spent on 'luxuries'. Low levels promote deprivation.

> **Key concept**
>
> **Inequality** is the outcome of uneven distributions. In this topic, the focus is on the uneven economic and social distributions that exist within societies and communities – for example, the uneven distributions of income and wealth, and of quality of life and social opportunities. The factors responsible for such differences are complex and resistant to remedy. One of the main aims behind much regeneration is to reduce such inequalities.

Income inequalities

Inequalities in pay levels are linked to differences in the type of employment. Some types of work (in the professions, for example) are more highly paid than others (such as manual work).

There are huge disparities in incomes and costs of living, both nationally and locally. This has always been the case, but the view is that these inequalities are increasing.

Quality of life correlates closely with wage and salary levels. This reflects the fact that many of the things that contribute to the overall quality of life are goods (e.g. housing and household equipment) and services (transport and leisure) that have to be paid for.

> **Quality of life:** The level of social and economic well-being experienced by individuals and communities, measured by various indicators such as health, longevity, happiness and educational achievement (Table 4.1).

Table 4.1 Factors affecting quality of life and inequality

Factors and processes	Possible measures
Economic inequality Employment opportunities, type of work and income	Employment/unemployment rates and types
	Average incomes
	Purchasing power (shopping basket surveys)
	Pound shop and loan shop surveys
	Land-use surveys
Social inequality Segregation of people and marginalisation or exclusion of subgroups	Age, gender, health, longevity, disability and educational achievement data from ONS and health authorities
	www.police.uk and crime and design surveys
	ACORN, CAMEO, Zoopla's ZED Index
	Village/community centre activities, e.g. playgroups, elderly social groups, internet blogs and email lists
	Placecheck
Service inequality Health facilities, public transport; food may be unequally available and accessed	Functional surveys of services (high/medium/low order)
	Bus timetable survey
	Taxi and community-shared transport surveys
	Supermarket and shop location survey
Environmental inequality Pollution levels, derelict land and access to open space have impacts on people's well-being	ONS: central heating provision
	Building quality surveys
	Environmental quality surveys on pollution, amount of and access to parks and green space, graffiti, derelict land, litter

Now test yourself

TESTED ☐

3 Distinguish between social and service inequalities.

Answer on p. 218

Typical mistake

It is erroneous to think that high levels of inequality exist only in developing and emerging countries. In fact, strong inequalities prevail in most so-called prosperous societies.

Revision activity

Make sure you know some of the possible measures of inequality shown in Table 4.1.

Now test yourself

4 Which economic factor most affects the quality of life? Justify your choice.

Answer on p. 218

Changing functions and characteristics

Functional and demographic changes

Over time, the functions of places change and therefore their economies change also. Traditionally, urban places have been involved in one or more of four key functions: administration, commerce, industry and services. Historically, high-order functions (such as banks, department stores, council offices and doctors' surgeries) have been located in larger settlements while lower-order functions (such as grocery stores, post offices and pubs) have been found even in small settlements.

However, commercial functions are now rapidly changing because of internet and broadband services and changing customer habits. The retail landscape has changed enormously with the advent of online shopping and banking, as well as click-and-collect.

While places change what they do for a living, so too do their populations. In many instances, it is the former conditioning the latter. Typical demographic changes include:
- trends – increasing or decreasing
- rates of change
- increasing ethnicity
- age and gender balances
- socio-economic structure – changing in response to processes such as **gentrification**, deindustrialisation, **deprivation** and **studentification**.

> **Gentrification:** The movement of middle-class people back into rundown inner-urban areas, resulting in an improvement of the housing stock and image.
>
> **Deprivation:** A condition when a person's well-being falls below a level generally regarded as a reasonable minimum. Measuring deprivation usually relies on indicators relating to employment, housing, health and education.
>
> **Studentification:** Social, economic and environmental change brought about by the concentration of students in particular areas and cities, usually located close to universities.

Measuring change

Changes within places can be measured in a number of ways, such as:
- land-use conversions
- employment trends
- demographic changes
- levels of deprivation.

The last of these is particularly important and much use is made of the index of multiple deprivation (IMD).

Now test yourself

5 Which employment trends would best show how a place has changed?

Answer on p. 218

Places and their connections

This part of your revision focuses on the two place studies you have completed as part of this topic. Figure 4.1 shows the main components of place, some of which you should have investigated in each of your contrasting places.

The specification recommends that you pay particular attention to:
● the regional and national connections (linkages) of your places
● the international and global connections of your places.

To what extent have these connections brought about change in your places? Have those changes impacted on local people?

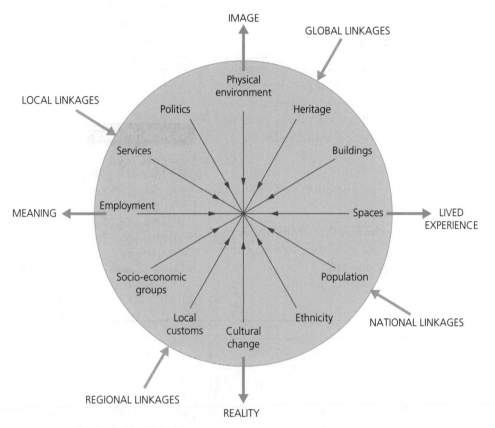

Figure 4.1 Components of place

Synoptic theme

Places in the UK, as elsewhere in the world, are being increasingly influenced by two major players: TNCs and IGOs. This is part and parcel of globalisation.

Your place investigations will have been guided by four groups of questions:
● those that establish the initial identities of your chosen places
● those about how their economic and social characteristics have been shaped by regional and national connections
● those about how their economic and social characteristics have been shaped by international and global connections
● those about how economic and social change have influenced the identities of people living in those places.

Synoptic theme

Attitudes towards place changes can be highly polarised between those that see them as eroding heritage and those that see them as place enriching.

Revision activity

It is recommended that you summarise your place studies by creating a table of two columns and four rows. There should be one column for each place and one row for each of the four groups of questions: place identity, regional and national influences, international and global influences, and people's identities.

Now test yourself

TESTED

6 To what extent have the connections of your two places changed?

Answer on p. 218

The need for regeneration

First, a reminder about the nature of regeneration. It is first and foremost an exercise in economic improvement. The hope is that the improvement will attract inward investment and create jobs. From this, hopefully there will be a spin-off of social benefits, such as a reduction in poverty and deprivation. There are two particular challenges associated with regeneration:

- persuading people of the need for it
- agreeing what would be the most appropriate and effective form for that regeneration to take.

Both challenges are made more difficult by the fact that people differ in so many ways, particularly in their lived experiences, their attachment to a place, their perceptions and who they happen to be.

Inequalities and perceptions

REVISED

> **Key concepts**
>
> **Lived experience** is the accumulated experience of living in a particular place. This can have a profound impact on a person's **perceptions**, values and identity, as well as on their general development and outlook on the world.
>
> Place attachment is the emotional bond between person and place. It is highly influenced by a person's lived experience and how long they have resided there.

The economic and social inequalities that lie rooted in employment and different levels of income affect people's perceptions of places. Place has a huge impact on us all, with the **lived experience** it has to offer and a sense of security that comes with feelings of attachment.

In addition, our perceptions of places are often coloured by the current fortunes of places.

Successful places

Places perceived as successful tend to be characterised by:

- high rates of employment
- high rates of in-migration (domestic and international)
- low levels of deprivation.

Such places become self-sustaining as more people and investment are drawn to the opportunities created. However, there may be negative externalities:

- increased property prices
- congestion of roads and public transport
- overburdened services, such as education and healthcare.

> **Exam tip**
>
> Sense of place and the meaning of place are rather 'slippery' terms. There is very little difference between them. They are so close that they may be taken to mean the same thing.

> **Typical mistake**
>
> Planned regeneration does not take place only in towns and cities. Regeneration may be more common in urban areas, but much is now being undertaken in rural areas.

> **Perception:** The 'picture' or 'image' of reality held by a person or group of people resulting from their assessment of received information.

The perception of residents in such places may differ:

- Younger people in high-earning jobs will enjoy the fast pace of life and opportunities.
- Unskilled people, lower earners and the long-term unemployed will have more negative views.
- Retirees may view them as too busy and look to other, perhaps less successful, places offering a slower pace of life with pleasant climate, sheltered accommodation and good access to healthcare.

It has to be emphasised that we may all have different views on what constitutes a successful place. Would everyone agree that London or San Francisco are successful places?

Less successful places

Economic inequality and technological change breed less successful places. For each successful place, there is at least one failing. Examples of such places abound and vary in scale from a whole region, such as the Rust Belt in the USA, through whole towns (for example, Hartlepool in northeast England) to a small inner-city area, say in Salford.

The symptoms of less successful places are all too obvious:

- high levels of poverty and deprivation
- unemployment
- derelict buildings
- graffiti
- crime and vandalism.

There is no doubt that less successful places are generally going to be perceived less favourably. The lived experience may be found wanting and the feelings of place attachment rather weak.

The particular challenge facing less successful places is becoming drawn into the spiral of decline (Figure 4.2).

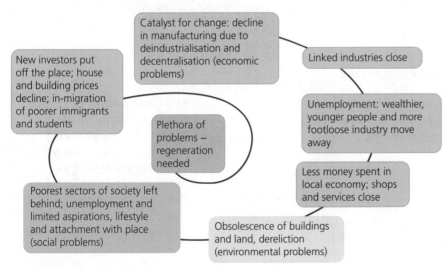

Figure 4.2 The spiral of decline

Priorities for regeneration

Economic and social inequalities create a need for regeneration, if only to target those places that are the victims of inequality, such as **sink estates**, areas left derelict by deindustrialisation and declining villages.

Revision activity

Make notes about a successful place. What factors have made it successful? Are there negatives?

Exam tip

Remember that not all unsuccessful places are urban. Think of the remote rural areas of the UK and those villages that were once prosperous farming communities.

Revision activity

Be sure you understand the sequence of linked events in the spiral of decline (Figure 4.2).

Sink estates: Housing estates characterised by high levels of poverty, deprivation and crime, such as domestic violence, drugs and gang warfare.

Now test yourself

7 Suggest three priority area for regeneration.

Answer on p. 218

The lived experience of place and engagement

Levels of engagement

One aspect of the lived experience that affects place attachment is the level of engagement. This is the degree to which a person participates in their local community; the degree to which they feel they belong to a place. It might be seen as a reflection of their place attachment. Indicators of engagement include:

● voting in local and national elections
● membership of, and participation in, local societies
● having a circle of local friends.

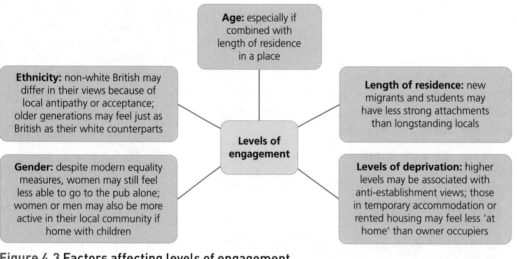

Figure 4.3 Factors affecting levels of engagement

The opposite condition to engagement is one of **marginalisation** and exclusion. The factors shown in Figure 4.3 as affecting levels of engagement can and do work in a negative way. People can feel excluded by ethnicity, gender and deprivation.

> **Marginalisation:** The social process of being made to feel apart or excluded from the rest of society. This leads to feelings of belonging to an underclass that is discriminated against.

Now test yourself

TESTED

8 What makes people feel marginalised and excluded?

Answer on p. 218

Typical mistake

Marginalisation and exclusion occur only in less successful places – not true!

Exam practice answers and quick quizzes at **www.hoddereducation.co.uk/myrevisionnotes**

Lived experience and place attachment

Lived experience and place attachment clearly vary from person to person. The factors shown in Figure 4.3 with respect to engagement also apply to lived experience and place attachment.

Conflicts

Conflicts often occur among contrasting groups in a community, largely because they hold different views about the priorities and strategies for regeneration. These can be caused by:

● a lack of political engagement and representation
● ethnic tensions
● social inequality
● lack of economic opportunities.

> **Synoptic theme**
> Players and whole communities may differ in their attitudes to urban regeneration and favour contrasting approaches. Those attitudes are influenced by the degree of attachment.

Now test yourself   TESTED

9 What factors affect a person's place attachment?

Answer on p. 218

Evaluating the need for regeneration   REVISED

It is quite possible for people living in the same place to have conflicting opinions even about whether there is a need for regeneration. To what extent was this the situation in your two chosen places?

Figure 4.4 sets out some of the criteria that are widely used when evaluating the need for regeneration.

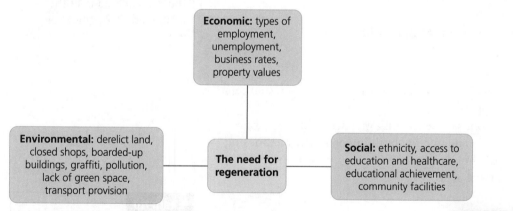

Figure 4.4 Some criteria for assessing the need for regeneration

In investigating the need for regeneration in your two places, you should have:

● used statistical evidence to determine the need for regeneration
● used different media to question the need for regeneration
● identified the factors influencing the perceived need for regeneration.

> **Exam tip**
> Remember that one of the most powerful criteria is unemployment. Note that most of the social and environmental criteria are spin-offs from this.

How regeneration is managed

The role of national government   REVISED

Investment in infrastructure

In the UK, government plays a key role in regeneration largely through investment in the national transport **infrastructure**. The thinking is that improved accessibility is the key to successful regeneration. Not only is it likely that investment in regeneration will be attracted to places with improved accessibility, but good accessibility will also be crucial in sustaining the regeneration. A current example is the HS2 project, a new high-speed rail link that will hopefully help to regenerate large parts of northern England.

Factors affecting regeneration policy are shown in Figure 4.5.

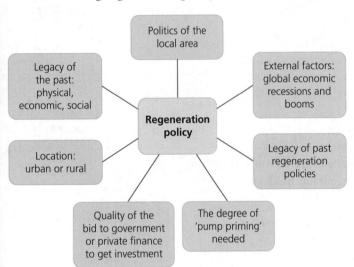

Figure 4.5 Factors affecting regeneration policy

Domestic policies

Government domestic policies can also help stimulate regeneration in a variety of ways, for example by:
- relaxing planning laws on, say, developing greenfield sites
- providing incentives to encourage the building of affordable housing
- allowing fracking in the hope that it might play a part in the regeneration of some rural areas.

International policies

The government can also pursue policies at an international level that have a direct bearing on regeneration. For example:
- deregulating capital markets to encourage foreign and private investment in regeneration schemes
- immigration: there is a tension here between the job-generating focus of regeneration and the availability of immigrant workers.

Now test yourself

TESTED

10 How does the UK government play a key role in regeneration?

Answer on p. 218

The role of local government

REVISED

Local plans

The main task facing local authorities is to create a sympathetic business environment that will support the regeneration. One of the most obvious ways is to have local plans that clearly designate areas for development or redevelopment, for example as retail, science or industrial parks. This is perhaps one of the most crucial decisions that a local authority has to make. What should be the lead activity heading up the regeneration – retailing, heritage and tourism, sport, or leisure and recreation? There is a range of regeneration options from which to choose (Figure 4.6). In some cases, the choice will be conditioned by the history of the proposed regeneration area.

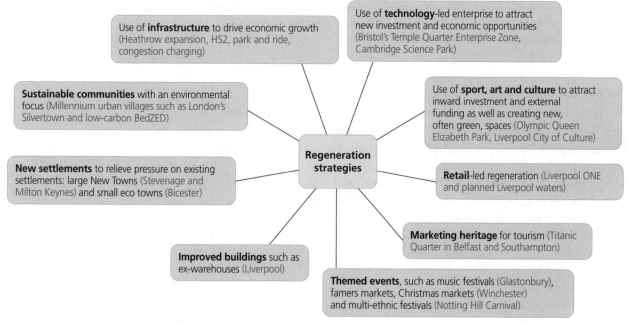

Figure 4.6 Regeneration strategies

Local support

Local governments will hope to receive local support for their regeneration schemes, as for example from the local Chamber of Commerce and the trade unions. However, everyone knows that such schemes will not be acceptable to everyone. Inevitably there will be tensions as some local interest groups express their support for preservation and conservation rather than redevelopment.

Now test yourself

TESTED

11 Describe ways in which local governments try to attract inward investment.

Answer on p. 218

Rebranding

REVISED

A new look

Using a variety of media and the processes of **rebranding** and **re-imaging**, attempts to make areas look attractive to potential investors, clients and local people involve changing both their outward appearance and public perception of them.

Rebranding deindustrialised places

Rebranding often focuses on the attractiveness of places. This is possible even in places bearing the scars of deindustrialisation either by comprehensive redevelopment and giving a place a totally new identity (for example, London Docklands) or by capitalising on a place's industrial heritage (for example, Telford). An added possibility in both instances is to use the regenerated places as venues for prestigious national and international events.

Rural rebranding strategies

The twenty-first century has seen a 'new rural economy' develop, with rural areas becoming much more like urban areas in their activities. A wide range of tried and tested strategies has emerged that seeks to rebrand this post-production countryside:
- capitalising on heritage and literary associations, for example Brontë country, where the Brontës lived and wrote their famous novels
- farm diversification – agriculturally (speciality products) and non-agriculturally (converting farm buildings into tourism accommodation or offices)
- outdoor sports
- adventure tourism
- counter-urbanisation – promoting telecottaging.

Now test yourself

TESTED

12 What evidence do you have of rebranding in your two place studies?

Answer on p. 218

Assessing the success of regeneration

> **Key concept**
>
> **Legacy** refers to the longer-term effects of regeneration schemes. It can be positive or negative. It might be judged on the re-use of landmark buildings, the amount of government support and private investment needed or whether local people benefit in the long term.

Possible measures

REVISED

The aim of regeneration is to create a **legacy** of increased employment and income, and reduced poverty and deprivation.

Economic measures

The term regeneration indicates a long-running process rather than a quick fix to economic, social and environmental problems, despite political and economic pressures for speed.

Possible economic measures are:
- employment: not just the scale of job growth but the mix of job types
- income: several possibilities here – average earnings, business turnovers, number of households on benefits
- poverty: a declining incidence and number of households on benefits.

These and other measures may be used to assess the performance of regeneration in two different ways:
- to compare the same measures both before and after regeneration
- to compare the results from one regeneration project with those of a similar project elsewhere.

> **Exam tip**
>
> Remember that 'success' can be subjective and that any improvement may not benefit all of the people living in an area.

Social progress

Social progress relates to how an individual and community improve their relative status in society. It can be measured by:
- reductions in inequalities both between and within areas
- reductions in the incidence of deprivation
- improvements in access to education and healthcare.

Quality of the living environment

Regeneration targeted at the built environment attracts more people to live there. Hopefully, it will also have positive knock-on effects on health. General improvements in aesthetics, security and safety are often common components of regeneration programmes. The effectiveness of these improvements can be seen through lowering of pollution levels and reductions in the amount of derelict land.

One final and important point to be made here is this: as already stated, most regeneration projects are not intended as quick fixes. For this reason, assessments of success or otherwise should not be attempted before the regeneration has had time to bed down and to reveal its strengths and weaknesses. Such a time lapse will be years rather than months.

Now test yourself

13 Check that you understand the three main measures of regeneration success and provide examples of each.

Answer on p. 218

Evaluation by urban stakeholders

Stakeholders generally

In the remainder of this and the next section, the specification expects you to take two regeneration schemes (one urban and one rural) and to focus on the following:

- the contested nature of the regeneration proposals – **stakeholders** for and against, and their particular views
- the impact of national and local policies and strategies in determining the nature of the regeneration
- the evaluation of the regeneration outcomes by different stakeholders.

Making judgements about the success of regeneration in any place involves not just those of the actual decision makers but also those of stakeholders – the people, groups and organisations with an interest in the regeneration. In most situations, stakeholders fall into four groups:

- providers: could be landowners, investors, contractors
- users or beneficiaries: those who stand to benefit (or lose out)
- governance: local government officials, enforcers of local bye-laws and national government policy
- influencers: action groups, political parties.

> **Stakeholder:** An individual, group or organisation with a particular interest in the actions and outcomes of a project or issue-solving exercise.

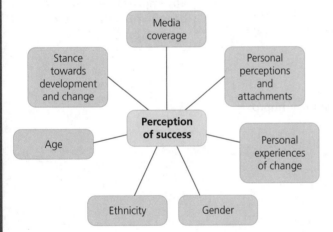

Figure 4.7 Factors influencing perception of success

From this, it follows that each stakeholder will have their own particular perception of, or opinion on, what constitutes 'success' and 'failure' (Figure 4.7). They will have their own vested interests and agendas. They will have their own criteria for assessing whether a particular scheme has been, or is being, managed successfully or not. Each stakeholder will arrive at its own verdict. Table 4.2 identifies the factors likely to influence the viewpoints and different urban stakeholders.

Table 4.2 Viewpoints and roles of different urban stakeholders

	Viewpoints	Roles
National government and planners	Reconciling different interests Longer-term national goals take priority	Planning permission Pump priming to start large, nationally important developments
Local councils	Have a duty to tackle inequality in their communities Make local planning decisions Are supposed to balance out the economic, social and environmental needs of a locality	Small or local regeneration schemes 'Soft management' helping regeneration, e.g. 'alcohol-free zones' Permissive arrangements, e.g. roller skating, street performers and artists
Developers	Economic standpoint: profit is needed	Funding of schemes
Local businesses	Views may be polarised: those expecting an increased customer base by the spin-offs from regeneration will differ from those threatened by it The local Chamber of Commerce may give majority viewpoints of business leaders	Lobby councils Invest in schemes
Local communities	The silent majority may be represented by a few willing and able to give up their time to be involved in either the local council or a pressure group; Broadwater Farm is an example of many players working together	Lobby councils Vote for local and national political parties Form pressure groups

Now test yourself

TESTED ☐

14 Check that you understand the difference between the four types of stakeholder shown in Table 4.2.

Answer on p. 218

> **Typical mistake**
>
> It is wrong to think that all stakeholders are equal when it comes to influencing the character and management of a regeneration scheme.

Stakeholders in an urban regeneration project

A possible urban case study for use in this part of the Specification might be Salford Quays, London's Olympic Park or Liverpool Waters.

> **Synoptic theme**
>
> Attitudes towards any regeneration proposal are more likely to be positive, but in all places NIMBYism will find a voice and press for no change here but somewhere else.

> **Revision activity**
>
> You do need to prepare notes on your chosen urban regeneration scheme under the three headings: contested nature (i.e. conflicts between stakeholders); impact of national and local policies; evaluation by different stakeholders.

Evaluation by rural stakeholders

REVISED ☐

Stakeholders in a rural regeneration project

A possible rural case study for use in this part of the specification might be Powys regeneration partnership, the North Antrim coast or Skye.

Figure 4.8 The Egan wheel

Figure 4.8 shows the Egan wheel, which creates an evaluative scoring system that can be used when assessing the outcome of regeneration projects in rural settings.

Exam tip

Remember that not all stakeholders are equal when it comes to influencing the management of an issue. When studying any issue, try to rank stakeholders according to the strength of their influence.

Now test yourself

TESTED

15 Check that you are able to provide examples of stakeholders and their views drawn from your own place investigations.

Answer on p. 218

Skills reminder

You should be familiar with the skills and techniques used to investigate the regeneration of places:

- use GIS to represent data about place characteristics
- interpret oral accounts relating to lived experience
- use the index of multiple deprivation to investigate spatial variations in deprivation
- use social media to understand how people relate to the places where they live
- test the strength of relationships by scatter graphs and Spearman's rank correlation
- investigate newspaper sources to understand conflicting local views
- evaluate different sources used in conveying the image of a place
- use media sources to explore how a place's identity has been changed by regeneration
- use photographic and map evidence to depict 'before' and 'after' scenarios
- interrogate blogs and other social media to understand different opinions on the performance of a regeneration project.

Exam practice

AS

1 (a) In which employment sector is retail? (1)

 (b) Study Figure 1.

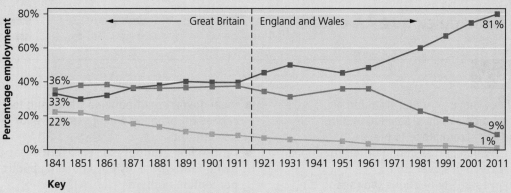

Key

— Primary sector (agriculture, forestry, fishing)

— Secondary sector (manufacturing)

— Tertiary sector (services, quaternary and quinary)

Figure 1 Sectoral change in the UK over 170 years

Source: Adapted from data from the Office for National Statistics, licensed under the Open Government Licence v.3.0

 (i) By how much did the tertiary sector grow between 1841 and 2011? (1)

 (ii) Calculate the difference between the tertiary and secondary sectors in 2011. (1)

 (iii) Suggest **one** reason for the decline in the secondary sector. (3)

 (c) Explain how economic restructuring has triggered a spiral of decline in some places. (4)

 (d) Explain the role that UK government policy plays in regeneration. (6)

 (e) Assess the extent to which rebranding attempts to represent areas as being more attractive. (12)

A-level

2 (a) Study Figure 2.

Figure 2 HS2, Heathrow and the Northern Powerhouse

(i) Suggest how HS2 will benefit people living in Birmingham. (3)

(ii) Suggest reasons as to how infrastructure development regenerates regions in the UK. (6)

(b) Explain why different urban stakeholders have different criteria for judging the success of urban regeneration. (6)

(c) Assess the extent to which the success of regeneration can lead to social progress and an improvement in the living environment. (20)

Answers and quick quiz 4A online

ONLINE

Summary

You should now have an understanding of:
- how economies can be classified in different ways and vary from place to place
- the way in which places change their functions and characteristics over time
- past and present connections that shape the economic and social characteristics of places
- economic and social inequalities that affect people's perceptions of an area
- the significant variations in the lived experience of places and levels of engagement with them
- a range of ways to assess the need for regeneration
- UK government policy decisions which play a key role in regeneration

- local government policies which aim to represent areas as being attractive for inward investment
- rebranding attempts to represent areas as being more attractive by changing public perception of them
- how a range of measures is used to assess the success of regeneration: economic, demographic, social and environmental
- that fact that different urban stakeholders have different criteria for judging the success of urban regeneration
- how different rural stakeholders have different criteria for judging the success of rural regeneration.

Option B Diverse places

There are four key concepts running through the whole of this topic: **place**, **population structure**, the **rural–urban continuum** and **ethnicity**.

Key concepts

Place is a part of geographical space with a distinctive identity and character that is deeply felt by local inhabitants. The sense of each and every place derives from a unique mix of external connections, natural and human features in the landscape and the people who happen to occupy it.

Population structure is the make-up of a population in terms of different age groups, the balances between those groups and between the sexes within them. Other components include life expectancy, family, size, marital status and ethnicity.

Rural–urban continuum is the unbroken transition from sparsely populated or unpopulated, remote rural places to densely populated, intensively used urban places (town or city centres).

Ethnicity is the cultural heritage shared by a group of people that sets them apart from others. The most common characteristics are racial ancestry, language, religion and forms of dress. It is an important component of cultural diversity in the UK.

Now test yourself

TESTED ☐

16 What is ethnicity and why is it such an important issue in many UK places?

Answer on p. 218

Population structure, time and place

Population structure

REVISED ☐

Population structure refers to the make-up of a population, particularly in terms of age and gender. Population structures vary from place to place and over time, as do population distributions. Both variations are particularly noticeable along the **rural–urban continuum**.

Population density

The place-to-place variations in populations are best shown in terms of **population density**. In the UK densities range from zero to more than 5,000 persons per square kilometre (Figure 4.9). There are many factors affecting population density, the most significant of which include:

- the physical environment, particularly relief and climate
- the economy, agricultural or non-agricultural, with the latter generating higher densities
- population characteristics – youthful structure is likely to raise densities
- planning, by controlling the location and amount of new housing.

> **Population density:** The number of people per unit area (usually per sq km), i.e. the total population of a given area (a country, region, city or place) divided by its area.

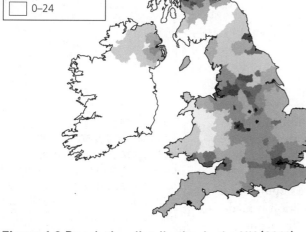

Key
Population density
(persons per km²)

⬛	5000 +
⬛	2500–4999
⬛	1000–2499
⬛	500–999
⬛	250–499
⬛	100–249
⬛	50–99
⬜	25–49
⬜	0–24

Figure 4.9 Population distribution in the UK (2011)

TESTED

Now test yourself

17 Identify the main features of the distribution of population in the UK as shown in Figure 4.9.

Answer on p. 218

Population dynamics

Over time, populations change, either increasing or decreasing. Change is the outcome of two processes: **natural change** and **net migration** (see Figure 4.10).

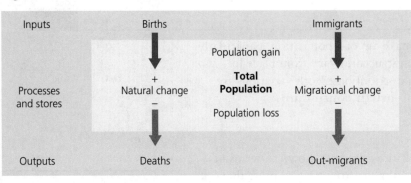

Figure 4.10 Components of population change

During the second half of the last century, population growth in the UK became very much concentrated in England, particularly in the South. The so-called 'North–South drift' occurred because of the rising economic prosperity of London and the Southeast as a global centre of finance and business, as well as a hub of modern service industries, and the decline of manufacturing (steel, shipbuilding, chemicals) in its former strongholds, namely the coalfields of the Midlands, the North, South Wales and central Scotland.

In the twenty-first century, it looks as if the UK has entered a new and different phase in its demographic history. There has been a slight evening out of population growth and population densities – a sort of rolling pin effect.

> **Exam tip**
>
> Be sure that you know the difference between density and distribution. Population density is one measure for showing the distribution of population.

> **Natural change:** The outcome of the balance between births and deaths in a population during a given period. Natural increase occurs when births exceed deaths; natural decrease occurs when deaths exceed births.
>
> **Net migration:** The balance at a national level between international arrivals (immigrants) and international departures (emigrants) during a given period. However, with areas within a country, the term takes into account both international and internal. The balances can be either positive or negative.

TESTED

Now test yourself

18 What are the two components of population change? Which do you think is responsible for more population growth in the UK?

Answer on p. 218

Population characteristics

REVISED

In terms of their impact on natural change in a population, age and gender are the two most important components. When these two are plotted in the form of a **population pyramid**, its detailed shape can tell us much about what has happened to the population over the last 75 or more years. It also provides some pointers as to how the population is likely to change in the foreseeable future. A basic distinction is made between youthful and ageing populations (Figure 4.11), the former promising population growth and the latter population stagnation or decline.

> **Population pyramid:** This is, in effect, a histogram, constructed in one-, five- or ten-year age groups, with males on one side and females on the other. The base of the pyramid represents the youngest age group and the apex the oldest.

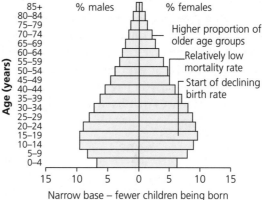

Figure 4.11 **A youthful and an ageing population**

Answer on p. 219

Now test yourself

TESTED

19 How might the recent influx of immigrants into the UK have impacted on the country's population pyramid?

Answer on p. 219

Age and gender are not the only components of population structure. Others include family and household size, marital status and ethnicity. These, and other demographic characteristics, such as life expectancy, vary between and within settlements. Within-settlement differences are well exemplified by the differences between the populations of places in the inner city and those of places in the suburbs. Compare, for example, the boroughs of Brent and Bromley in Greater London in terms of their age structures and ethnicity. Between-settlement differences are well shown along the rural–urban continuum. Compare, for example, the populations of commuter villages with those in remote rural places.

One very obvious demographic and cultural variable is **ethnicity**. Ethnic diversity is most marked in the inner city and generally declines along the rural–urban continuum.

The ethnicity and cultural diversity of places in the UK are being changed by:
- the impacts of migration
- the social clustering of immigrants
- the pull of major cities
- government planning policies.

> **Typical mistake**
>
> Population structure is not only about age and gender. There are other components (see below) but perhaps of less significance, with the notable exception of ethnicity.

> **Synoptic theme**
>
> The actions of governments may foster or suppress cultural diversity. The number of governments falling in the latter category is dwindling fast. It would be true to say that an increasing number of governments are neutral on this issue.

Now test yourself

TESTED

20 What is the significance of family size and marital status as components of population structure?

Answer on p. 219

Place character and connections

REVISED

This part of your revision focuses on the two place studies you have completed as part of this option topic. Figure 4.12 shows the components that you should have investigated in order to analyse the image (real and perceived) of each of your two contrasting places. The specification recommends that you pay particular attention to the external connections and influences that have affected both continuity and change.

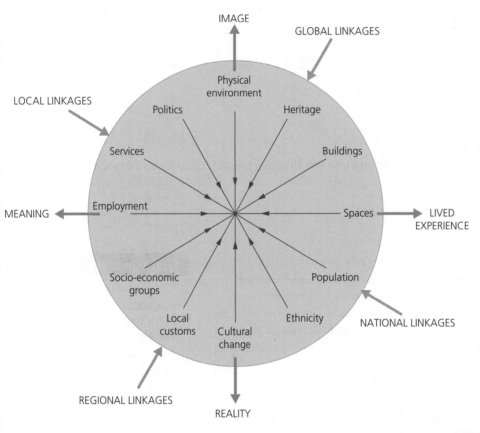

Figure 4.12 Components of place

Now test yourself

TESTED

21 Of the internal factors shown in Figure 4.14, which do you think have had the most impact on the identity of your two places?

Answer on p. 219

Living spaces

This and the remaining enquiry questions in this topic need a sound grasp of three key concepts (**lived experience**, **living space** and **perception**) that are essential to an understanding of this part of the topic, which is about the lived experiences derived from both urban and rural places.

Revision activity

It is recommended that you summarise your place studies by creating a table of two columns and four rows. You should have one column for each place and one row for each of the following aspects of your places: thumbnail portraits and initial impressions; main internal factors; external influences; and recent demographic and cultural change. Fill in each of the eight slots of your table with the appropriate summary notes.

The specification suggests that you pay particular attention to the impact of external influences, such as TNCs and IGOs.

Exam tip

Sense of place and meaning of place are rather 'slippery' terms. There is very little difference between them. They are so close that they may be taken to mean the same thing.

Perception: The 'picture' or 'image' of reality held by a person or group of people resulting from their assessment of received information.

Urban living spaces

REVISED

Popular perceptions of urban places change over time. The industrial towns and cities created by the Industrial Revolution were viewed, with the exception of their employment opportunities, in a generally negative manner, largely because of the poor quality of housing and the appalling pollution.

The lived experience afforded by urban places today is altogether better. Perceived pluses include:
- employment and career opportunities
- relatively high wage and salary levels
- quality and accessibility of social and commercial services
- leisure amenities and entertainment.

But the 'bright lights of the city' are dimmed rather by:
- the high cost of living
- sky-rocketing house prices relative to wage and salary levels (see Figure 4.13)
- low environmental quality – substandard housing, pollution in various forms
- social isolation of the elderly and members of ethnic groups
- crime.

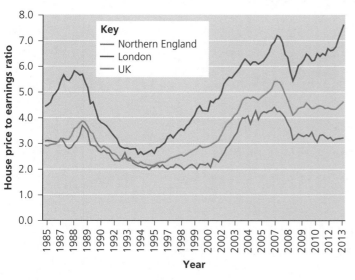

Figure 4.13 House price to earnings ratios, 1985–2013

Today, many people prefer the lived experience of the suburbs to that of inner-urban places. The suburbs are perceived as better for fulfilling family needs, such as offering:

- more residential space
- better schools and healthcare
- easy access to shops meeting daily needs
- more amenity green space.

But even in the suburbs, the rising cost of housing and other negative aspects are driving people out of the city altogether. Commuting from dormitory villages has become a preferred lifestyle.

What are perceived and identified as the pros and cons of the urban lived experience will depend on who you are, your changing values and attitudes, and where you are in the life cycle.

Now test yourself

TESTED

22 Identify the advantages of living in inner-urban areas.
23 Suggest a population group that is likely to find the inner-urban living space appealing.

Answers on p. 219

Rural living spaces

REVISED

Just as there are different living spaces in urban areas, so there are also in the range of rural places spread out along the rural–urban continuum. There are three main types of rural living space:

- Commuter belt: popular with adults of working age with children. Accessibility to urban jobs is the key, plus the perception that here people are able to escape from the downside of urban places. There are fast rates of population growth.
- Accessible rural: popular with retired people and with urban-based day-trippers. Population structure is unlikely to be too disturbed by domestic or international migration.
- Remote rural: victims of depopulation and the spiral of decline in such areas (Figure 4.14). Recently, some places have experienced a reversal of fortunes, thanks to counter-urbanisation, the communications revolution and tourism.

People, especially the young, leave for better opportunities in urban places

Employers find it difficult to recruit labour

Reduced investment in the area and businesses close

Less money, less employment and fewer people lead to shops closing and services declining

People become more aware of the general decline and the lower quality of life

Figure 4.14 The spiral of decline in remote rural places

These different rural places may also be distinguished demographically:
- the 'young' populations of the commuter belt
- the 'normal' populations of accessible rural areas
- the ageing populations of remote rural areas. In most rural places, levels of ethnicity are much lower than in urban places.

Part of the attraction drawing people to rural living spaces is their pursuit of the **rural idyll**. This seems to be the case particularly with urban-to-rural retirement moves. But the reality is that the lived experience in rural places often has a downside. This includes:
- declining public transport and difficulties of access to town-based services
- isolation and loneliness – distance from friends and relatives back in town
- tensions between incomers and longstanding residents
- communications – poor broadband service, 'no signal' areas in the mobile phone networks.

> **Rural idyll:** A 'chocolate box' image of quaint villages set in beautiful countryside. A place thought to be free of most of the negatives associated with urban living.

Synoptic theme

Attitudes towards suburban and inner-city living not only change over time as part of changing tastes and fashion, they also change as people progress through the life cycle.

Revision activity

Refer to your notes of how the nineteenth-century novelist Thomas Hardy portrayed the rural idyll in his fictional county of Wessex. Was it a completely 'dream county'?

Typical mistake

Do not assume that everyone sees the rural idyll in exactly the same way. One person's dream can be another's nightmare.

Now test yourself

TESTED

24 Describe the spiral of decline experienced in remote rural areas.

Answer on p. 219

Evaluating perceptions of living spaces

REVISED

It is quite possible for people living in the same place to have conflicting perceptions and different lived experiences. To what extent is this the situation in your two chosen places? Investigating this issue is probably best done by means of interviews and questionnaires, researching how your places are viewed by the media, and making the link between particular perceptions and lived experiences.

Another investigation required by the specification is that of demographic and cultural change. Demographic change relates mainly to total numbers and age structures, while cultural change relates to shifts in the ethnic mix. Are either or both of the changes giving rise to tensions? But there is also physical evidence of change, such as the upgrading or deterioration of housing, and the expansion or decline of social and commercial services.

> **Synoptic theme**
>
> Urban and rural residents may differ in terms of their **attitudes** towards places and those attitudes may vary, depending on the type of place.

Now test yourself

TESTED

25 What evidence do you have that local residents hold different views about the place in which you live? Suggest reasons for the differences.

Answer on p. 219

> **Revision activity**
>
> Be sure you have produced notes summarising the results of your two place investigations, as suggested on page 102.

> **Exam tip**
>
> Provided the evidence is relevant, you are recommended to use material from your own place studies to illustrate and support answers to more general questions.

Tensions in diverse places

Migration and diversity

REVISED

Internal

Migration (internal and international) is an important component of population change. For much of the twentieth century, a major internal migration within the UK was the so-called 'North–South drift' – the general movement of people from the northern parts of the country to the Southeast, and to London in particular. At the same time, particularly in the South, there were two decentralising migrations, namely from inner-urban areas to the suburbs and to more distant commuter dormitories. At the turn of the millennium, the character of decentralising migration began to change with the onset of **counter-urbanisation** and growing volumes of urban-to-rural retirement moves.

Now test yourself

TESTED

26 What is counter-urbanisation?

Answer on p. 219

Remember that:
- internal migration within a country is a 'zero sum' phenomenon in that any net gain in one area can occur only if there is a net loss of migrants elsewhere within the country
- migration is two-way traffic
- one outcome of migration is a mixing of the population as the people moving into an area often can be quite different from those moving out
- this mixing can generate tensions, as for example between newcomers and longstanding residents.

> **Exam tip**
>
> The last three bullet points also apply to international migration, except that the mixing is more about ethnic and cultural groups.

International

An acute shortage of manpower following the Second World War encouraged migrant workers and their families to come to the UK from what were or had been colonies in Africa and the Caribbean, and from what had been the Indian Empire (now divided into India, Pakistan and Bangladesh). These immigrant flows converged on the major cities where job opportunities were most abundant. They have had a profound impact on the ethnicity and cultural diversity of the UK's population.

The volumes of immigration flows have varied over time according to the booms and slumps in the UK economy. Today immigration into the UK involves significant source areas within the EU, particularly from East European countries like Poland (see Figure 4.15). Jobs, relatively good levels of remuneration and possibly access to social benefits are the main pull factors.

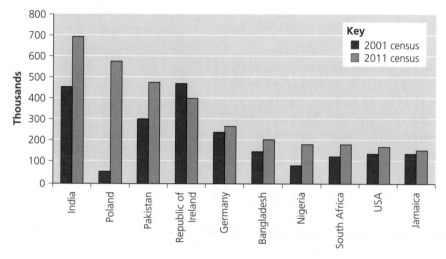

Figure 4.15 UK immigration: top 10 source countries (2001 and 2011)

The most recent immigrants still largely head for towns and cities. It is interesting to note that even those immigrants finding work in the fields of eastern England actually live in nearby towns, such as Boston (Lincolnshire) and Thetford (Norfolk).

Synoptic theme

In many countries, the government is the main player affecting migration flows.

Exam tip

Be clear as to the main factors attracting immigrants to the cities and towns of the UK. They are the availability of jobs and cheap housing as well as the snowball effect of choosing to be near to family and friends who have already come from the same source area.

Typical mistake

It is wrong to think that many migrants are choosing to live in rural areas. They may do rural work but they live in towns.

Revision activity

Check your notes about 1) domestic migration within the UK, and 2) UK immigration since 1950 (from Commonwealth countries and, more recently, from Eastern Europe). Highlight the pull factors.

Now test yourself

TESTED

27 How did the main sources of UK immigrants change during the last intercensal period?

Answer on p. 219

Segregation

REVISED

The 2011 Census showed that the white British element in the population of England and Wales had fallen to 80 per cent from a figure of 87 per cent in the previous census. But the distribution of immigrant ethnic groups is not an even one. In London, the white British element is now only 60 per cent of the total population; in all the other regions (with the exception of the Southwest, Northeast and Wales) the figure is close to 80 per cent. In all regions, the ethnic minority groups are most concentrated in towns and cities, and within those settlements still further segregated into **enclaves**.

> **Enclave:** A group of people surrounded by a group or groups of entirely different people in terms of ethnicity, culture or wealth.

There are two contradictory views of the possible reasons for this spatial segregation of minority groups. One argues that it is encouraged by external factors, such as the availability of cheap housing, with the host population 'forcing' the segregation (Figure 4.16). The other view emphasises internal factors, such protection and mutual security, that see segregation as being the wish of the segregated groups.

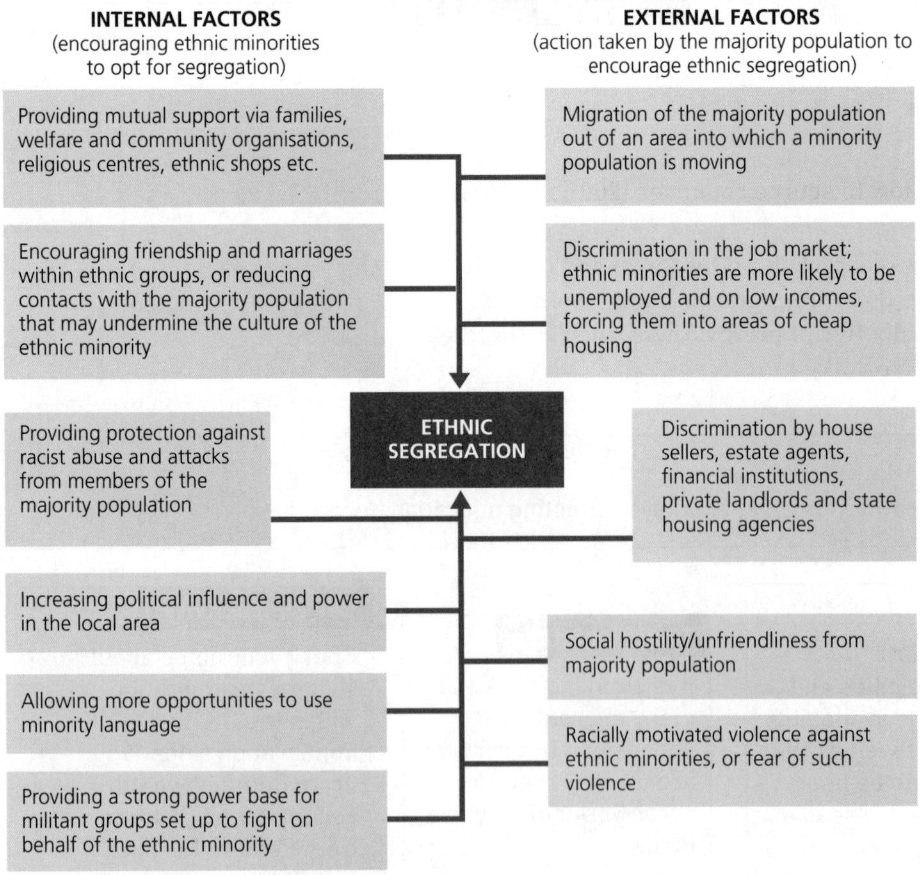

INTERNAL FACTORS
(encouraging ethnic minorities to opt for segregation)

EXTERNAL FACTORS
(action taken by the majority population to encourage ethnic segregation)

Providing mutual support via families, welfare and community organisations, religious centres, ethnic shops etc.

Migration of the majority population out of an area into which a minority population is moving

Encouraging friendship and marriages within ethnic groups, or reducing contacts with the majority population that may undermine the culture of the ethnic minority

Discrimination in the job market; ethnic minorities are more likely to be unemployed and on low incomes, forcing them into areas of cheap housing

ETHNIC SEGREGATION

Providing protection against racist abuse and attacks from members of the majority population

Discrimination by house sellers, estate agents, financial institutions, private landlords and state housing agencies

Increasing political influence and power in the local area

Social hostility/unfriendliness from majority population

Allowing more opportunities to use minority language

Racially motivated violence against ethnic minorities, or fear of such violence

Providing a strong power base for militant groups set up to fight on behalf of the ethnic minority

Figure 4.16 Two views of the factors encouraging ethnic segregation

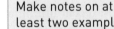

> **Revision activity**
>
> Make notes on at least two examples of ethnic segregation in the UK, paying particular attention to location and causal factors. Suitable London place studies include the ethnic concentration in the borough of Brent, the changing distribution of Jewish communities and the arrival of the Russian oligarchs and their families. The cities of Birmingham and Manchester also offer plenty of place examples.

Given that ethnic segregation has persisted in the UK for decades, if not centuries, it is hardly surprising that different groups should make their mark on the urban landscapes they occupy. The most conspicuous markers are places of worship (mosques, temples, synagogues, etc.), restaurants and food stores reflecting the culture and food preferences of the local population. It is said that the casual visitor to Southall (London) might be forgiven for thinking that they might be in part of Mumbai, Dhaka or Karachi, so great is the 'Indianisation' of the urban landscape.

An important aspect of segregated ethnic communities is what happens to them in the long term. It appears that there is an increasing degree of **assimilation** in successive generations and a moving out of the enclave. Such moves are encouraged by a variety of factors:

- improved earnings and better employment prospects
- feeling more confident and secure in the host society
- wishing to become more integrated into the host society.

Even if there is assimilation, the strength of the ethnic enclave is not diluted by the outward moves. This is because the people who move out are replaced by newly arrived immigrants belonging to the same ethnic group.

> **Assimilation:** The process by which different groups within a community intermingle and become more alike. The process particularly applies to the integration of immigrant ethnic groups.

Now test yourself

TESTED ☐

28 Suggest possible measures of assimilation.

Answer on p. 219

Change, tension and conflict

REVISED ☐

Change is so often the generator of tension, be it a change in land use, say from residential to commercial, or a change in the people occupying an area of housing. In fewer instances, change can be of such proportions as to lead to conflict. The issue is that in virtually all changing situations there are winners and losers, those who benefit and those who miss out. Any losers inevitably feel aggrieved and even 'cheated'.

The changing pattern of urban land use is the outcome of competition for space. This competition involves two distinct layers of living space:

- competition between housing and other consumers of space (services, commerce and industry)
- competition for housing space within residential areas.

> **Exam tip**
>
> For many, tension and conflict are synonymous terms. However, here it is advisable to distinguish between them, with tension seen as a precursor of conflict.

The fact that most urban areas are expanding is creating an insatiable demand for more space for housing and a range of urban activities. When it comes to bidding on the urban land market, there is a clear pecking order. Retailing and high-order offices are usually the strongest bidders, with housing and recreation at the bottom of the bidding league table.

Four housing sectors compete for space:

- owner-occupiers
- property developers acquiring housing to rent to tenants
- housing associations providing affordable housing
- local authorities providing social housing.

The arrival of immigrants has added to the pressure on an inadequate supply of housing and increased the competition for residential space. The emergence of ethnic minority enclaves in British cities has created a fair amount of ill-feeling on the part of white residents. So-called 'race riots' (e.g. Notting Hill 1958, Toxteth 1981, Bradford 2001) have been the expressions of mounting tensions. Similarly, the **gentrification** of residential areas has created, and still is creating, tensions between those incomers investing in the improvement of dwellings and the longstanding residents who feel they are being forced out of their homes.

Much change in urban areas is motivated by market forces. But a significant amount has also been the responsibility of government (national and local). Classic examples include the clearance of slums in the early postwar years and their replacement by high-rise blocks of flats. The lived experience of many occupying those blocks was found to be isolating and alienating. Equally, the apparent neglect by the public authorities of areas of acute poverty and deprivation leads to feelings of social exclusion and resentment, particularly among migrant groups (e.g. in parts of Glasgow).

> **Gentrification:** The movement of middle-class people into rundown, inner-urban areas and the associated improvement of the housing stock and area image.

Synoptic theme

Planners and developers may make controversial decisions. Their priorities and attitudes may differ from those of local groups.

> **Revision activity**
>
> Check through your notes for examples of the different tension scenarios. Possible place studies include Lewisham (London), Luton and Glasgow.

Now test yourself

TESTED

29 What evidence do you have of the existence of tensions in your two place studies? What appear to be causes of those tensions?

Answer on p. 219

Managing cultural and demographic issues

There are many aspects of change in both rural and urban places that need some form of **management** in order to achieve a satisfactory outcome. What constitutes an issue is important to an understanding of this final part of the topic. Figure 4.17 indicates some demographic and cultural issues that would qualify for management.

> **Management:** A set of actions that facilitates the transition from one situation to another. More specifically, those actions might be aimed at solving or ameliorating a particular problem or issue.

Key concept

An **issue** is an important topic or problem for discussion and hopefully solution by means of planning and management. Examples relevant to this topic include immigration, housing, discrimination and segregation, poverty and deprivation.

Ethnicity
- Assimilating ethnic minorities
- Respecting immigrant cultures
- Outlawing discrimination
- Conserving cultural heritage

Population structure
- Anticipating future change
- Encouraging a youthful population
- Coping with an ageing population
- Raising life expectancy

CULTURAL AND DEMOGRAPHIC ISSUES

Migration
- Reducing native versus incomer tensions
- Stemming unwanted outflows
- Controlling immigration
- Improving border security

Quality of life
- Improving access to, and quality of, housing
- Providing healthcare and education
- Reducing poverty and deprivation
- Improving the living environment

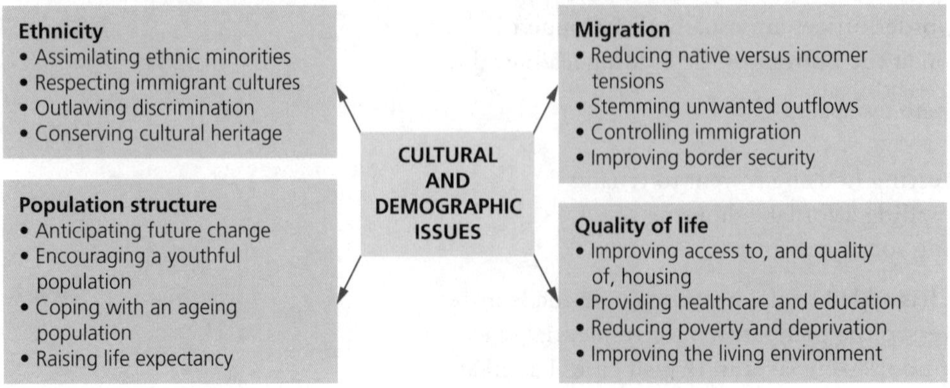

Figure 4.17 Possible cultural and demographic issues requiring management

Evaluation

A crucial stage is reached in the management of an **issue** when it becomes necessary to take stock of what has been done and to assess whether or not it has achieved or is achieving the original objective. There is a variety of potential measures available here. Which of these to use depends on the issue. Table 4.3 provides three examples.

Table 4.3 Some indicators or measures of issue management

Issue	Possible measures
Assimilation and integration	Number of people moving out from enclaves Voter turnout Incidence of hate crime Involvement in the wider community
Social progress	Trend in index of multiple deprivation (IMD) Unemployment rate Number of households on social benefits Change in life expectancy
Housing provision	Number of new units built Percentage breakdown – for sale and rent Incidence of social housing Council housing waiting list

Now test yourself

TESTED

30 Suggest another possible measure for each of the issues in Table 4.3.

Answer on p. 219

Stakeholders

REVISED

Making judgements about the management of change in any place involves not just the actual decision makers but also **stakeholders** – the individuals, groups and organisations with an interest in the issue to be managed. In most situations, stakeholders fall into four groups:

- **Providers**: could be landowners, investors, contractors.
- Users or beneficiaries: those who stand to benefit (or lose out).
- Governance: local government officials; enforcers of local bye laws and national government policy.
- Influencers: action groups, political parties.

Each stakeholder, no matter whether the issue relates to an urban or a rural place, will have its own vested interests, particular perceptions and agendas.

From this, it follows that each stakeholder will have its own particular view or opinion of what constitutes 'success' and 'failure'. It will have its own criteria for assessing whether a particular issue has been, or is being, managed successfully or not. Each stakeholder will arrive at its own verdict.

In the remaining two sections, the best that can be done is to take two issues (one urban and one rural) and identify the stakeholders as well as causal factors. In both, the emphasis is on the management of change.

> **Stakeholder:** An individual, group or organisation with a particular interest in the actions and outcomes of a project or issue-solving exercise.

> **Typical mistake**
>
> It is wrong to think that everyone is agreed about what constitutes a successful outcome in the management of urban and rural issues.

> **Typical mistake**
>
> It is wrong to believe that all stakeholders are equal when it comes to influencing the management of an issue.

Now test yourself

31 Check that you understand the difference between the four types of stakeholder.

Answer on p. 219

An urban issue

In meeting the rising demand for housing in a rapidly growing city surrounded by a green belt and where there is very little land for housing within its boundaries, the issue is where to build the houses once the available land within the city has been developed. Where outside the city should the required housing be built?

The stakeholders fall into two groups: internal and external. Table 4.4 identifies some of them and their particular interests and perceptions.

Table 4.4 **A sample of stakeholders**

Stakeholders	Particular interest
Internal stakeholders	
City council	Shortage of housing does not hinder the growth of the city's economy
Employers and businesses	Difficulties in recruiting labour as workers are deterred by the shortage and consequent expense of housing
Social service providers	The rising demand for education and healthcare; recruiting labour to the social services
Local residents	Impact of more housing on city traffic and quality of living space
External stakeholders	
National government	That everything is done to keep the city flourishing
County council	Anxious to provide the space needed for new housing, but at the same time keen not to alienate local people
Conservation groups	Protection of designated green belt and concern about the possible impact of housing developments on other rural places and wildlife
Local residents in areas designated for new housing	Maintaining the status quo and keeping rural areas rural; impact of new housing on property values

Synoptic theme

What is deemed to be 'successful' depends on the attitudes of different players.

Revision activity

The issue of improving the quality of urban life is only one of a number of urban issues that you might have studied. For the managed issue you have studied, identify the main stakeholders and set out your notes as in Table 4.4 or Figure 4.18. Other possible issue-oriented place studies include Southall (London) and Oxford.

A rural issue

How to maintain commercial and social services is a common issue in many rural parts of the UK. Another is how to deal with increasing pockets of deprivation and poverty. Figure 4.18 illustrates the views of internal and external stakeholders in one such problem area, Breckland in Norfolk. What can be done to improve the quality of life in this local authority area? To what extent do stakeholders' views converge to define a common course of action?

(a)

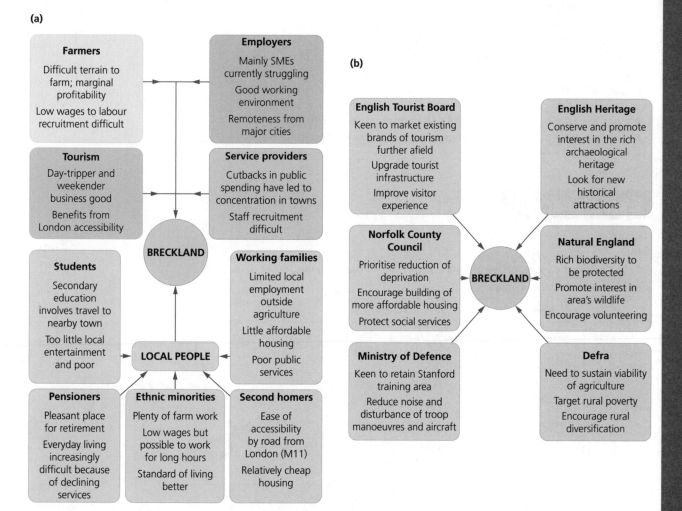

Figure 4.18 Breckland stakeholders: (a) internal, (b) external

---

Synoptic theme

What is deemed to be 'successful' depends on the attitudes of different players.

Revision activity

The above is only one of a number of rural issues you might have studied. For the managed issue you have studied, identify the main stakeholders and summarise your notes as in either Table 4.4 or Figure 4.18.

Exam tip

Always remember that not all stakeholders are equal when it comes to influencing the management of an issue. When studying any issue try to rank stakeholders according to the strength of their influence.

32 Check that you are able to provide examples of stakeholders drawn from your own place investigations.

Answer on p. 219

Skills reminder

You should be familiar with the skills and techniques used to investigate the following aspects of places:

- identity: investigating media images and various forms of qualitative data
- perceptions and lived experiences: interviews, oral accounts and social media
- spatial variations: using GIS; indices of ethnicity and cultural diversity
- change: interpreting old photographs and maps, newspaper reports.

Exam practice

AS

1 (a) What measure is commonly used to show the distribution of population? (1)

(b) Study Figure 1.

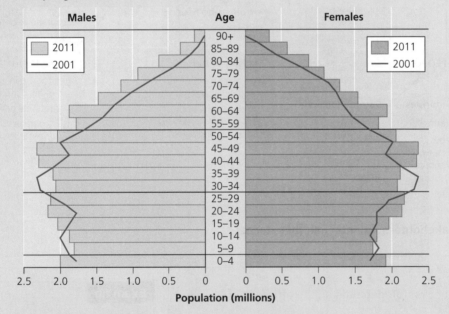

Figure 1 The UK's population pyramids, 2001 and 2011

(i) Describe **two** ways in which the population structure of the UK changed between 2001 and 2011. (2)

(ii) Suggest **one** reason for the bulge in the population pyramids between the ages of 30 and 55. (3)

(c) Explain the reasons why people retire from urban to rural places. (4)

(d) Explain what attracts immigrants to particular locations within urban places. (6)

(f) Assess the value of social media and blogs when investigating different perceptions of places. (12)

A-level

2 (a) Study Figure 2.

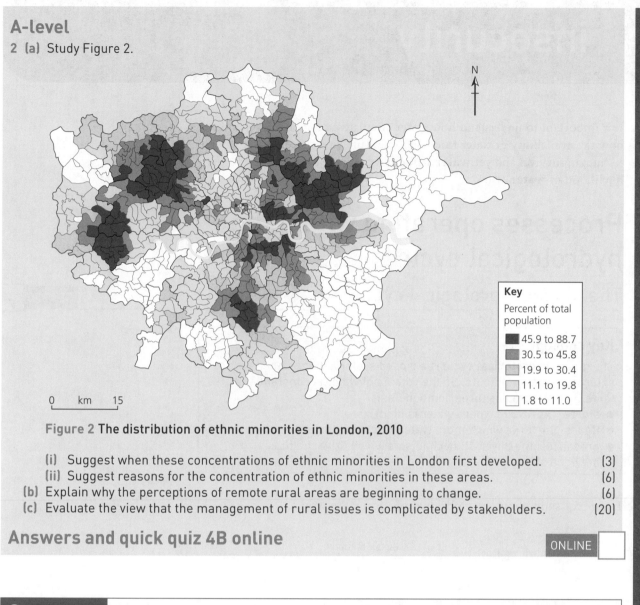

Key

Percent of total
population

- ■ 45.9 to 88.7
- ▨ 30.5 to 45.8
- ▨ 19.9 to 30.4
- ▢ 11.1 to 19.8
- □ 1.8 to 11.0

0   km   15

Figure 2 The distribution of ethnic minorities in London, 2010

(i) Suggest when these concentrations of ethnic minorities in London first developed. (3)
(ii) Suggest reasons for the concentration of ethnic minorities in these areas. (6)
(b) Explain why the perceptions of remote rural areas are beginning to change. (6)
(c) Evaluate the view that the management of rural issues is complicated by stakeholders. (20)

Answers and quick quiz 4B online

ONLINE

Summary

You should now have an understanding of:
- how population numbers, densities and structures vary over time and from place to place, particularly along the rural–urban continuum
- the meaning of place and the factors that give places and their inhabitants a sense of identity
- both urban places and rural places as seen differently by different groups; this is the outcome of variations in lived experiences and perceptions of places

- society and culture in the UK increasing in diversity largely as a result of immigration, first from Commonwealth countries and, more recently, from Eastern Europe
- segregation as a feature of immigrant ethnic groups in the towns and cities of the UK
- demographic and cultural change leading to tension and even conflict
- demographic and cultural issues requiring management; any assessment of management depends on the issue as well as on the stakeholders and their criteria.

5 The water cycle and water insecurity

It is important to understand how water circulates at a global level and how the availability of water fluctuates over time and space. Water is vital to human survival and yet **water insecurity** is increasing and so too the likelihood of **water wars**.

Processes operating within the hydrological cycle

The global hydrological cycle

REVISED

> **Key concept**
>
> The **global hydrological cycle** is a closed system. It does not have any external inputs or outputs. So the volume of water is constant and finite. The system has three components:
> - stores: reservoirs where water is held
> - fluxes: the flows which move water between stores
> - processes: the physical mechanisms which drive the fluxes between stores.

Figure 5.1 shows how the global hydrological cycle works.

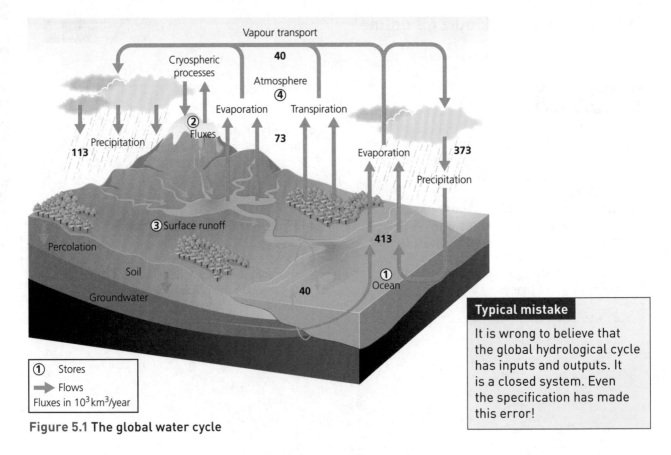

Figure 5.1 The global water cycle

> **Typical mistake**
>
> It is wrong to believe that the global hydrological cycle has inputs and outputs. It is a closed system. Even the specification has made this error!

Exam practice answers and quick quizzes at **www.hoddereducation.co.uk/myrevisionnotes**

The energy motivating the global hydrological cycle comes from two sources: the Sun and gravity.

Stores

The main stores of water are the oceans (96.9 per cent), ice caps and glaciers (1.9 per cent), groundwater (1.1 per cent), and rivers and lakes (0.01 per cent). When it comes to freshwater alone, the store situation is very different: ice caps and glaciers (68.7 per cent), groundwater (30.1 per cent), and rivers and lakes (1.2 per cent). With freshwater, a distinction is sometimes made between **blue water** and **green water**.

Water stores have different **residence times**. In general, the larger the store, the longer the residence time. Some stores are non-renewable; for example, **fossil water** and when the **cryosphere** melts.

Flows and processes

Flows are the transfers of water from one store to another. They are achieved by processes such as precipitation, evaporation, transpiration, cryospheric exchanges and runoff.

From a human viewpoint, the most important aspect of the hydrological cycle is the accessibility of freshwater. The global water budget limits this amount. The accessible sources are lakes (natural and artificial), rivers and groundwater. Overall only about 1 per cent of global freshwater is easily accessible for human use. But this figure is partly determined by the current level of technology. Presumably it might be increased in the future, but what might the impact be on the hydrological cycle as a whole?

Blue water: Freshwater stored in rivers, streams and lakes – the visible part of the hydrological cycle.

Green water: Freshwater stored in the soil and vegetation – the invisible part of the hydrological cycle.

Residence time: The average time a water molecule will spend in a store or reservoir.

Fossil water: Ancient, deep groundwater from former pluvial (wetter) climatic periods.

Cryosphere: Water frozen into ice and snow.

Exam tip

'Fluxes' and 'transfers' are two alternative terms for 'flows'. It is recommended that you stick to flows.

Now test yourself

TESTED

1 Why is only such a small percentage of freshwater accessible for human use?

Answer on p. 219

Exam tip

Remember that:
- climate change impacts on the hydrological cycle, most noticeably on the relative importance of the different stores
- the hydrological cycle is affected in a small way by human interventions, for example by water storage reservoirs and irrigation schemes.

The drainage basin as an open system

REVISED

The drainage basin is a subsystem within the global hydrological cycle. It is an open system with inputs and outputs. Drainage basins can vary enormously in size. They commonly 'nest' within each other. The drainage basin of a major river like the Amazon is made up of the drainage basins of tributary rivers. Those tributary basins, in turn, are made up of the even smaller basins of streams draining into those tributaries.

Inputs and outputs

The water inputs of a drainage basin are:

● precipitation
● flows from tributary basins.

The water outputs are:

● evaporation
● transpiration
● flows to a higher-order drainage basin or the sea.

Flows

The flows of water within a drainage basin are shown in Figure 5.2.

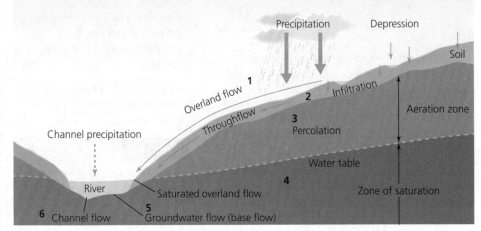

Figure 5.2 How the various flows operate within the drainage basin system

Physical factors influencing the drainage basin cycle

Of the factors in Table 5.1, climate is the most important because it determines the most important input, precipitation, as well as a significant output, evaporation.

Soils, geology and vegetation affect virtually all transfers, from infiltration and interception to throughflow and overland flow. Relief is significant through its impact on precipitation and runoff.

Table 5.1 The impact of physical factors within the drainage basin on inputs, flows and outputs

Climate	Climate has a role in influencing the type and amount of precipitation overall and the amount of evaporation, i.e. the major inputs and outputs. Climate also has an impact on the vegetation type.
Soils	Soils determine the amount of infiltration and throughflow and, indirectly, the type of vegetation.
Geology	Geology can impact on subsurface processes such as percolation and groundwater flow (and, therefore, on aquifers). Indirectly, geology alters soil formation.
Relief	Relief can impact on the amount of precipitation. Slopes can affect the amount of runoff.
Vegetation	The presence or absence of vegetation has a major impact on the amount of interception, infiltration and occurrence of overland flow, as well as on transpiration rates.

Now test yourself

TESTED

2 What are the direct and indirect impacts of geology on the drainage basin cycle?

Answer on p. 219

Human factors influencing the drainage basin system

Human activity can affect the drainage basin system, particularly the flows. Consider the impacts of deforestation on:

- evaporation – increased
- runoff – increased
- evapotranspiration – reduced
- interception – reduced
- infiltration – reduced
- groundwater – reduced.

Now test yourself

TESTED

3 Are you able to explain each of the bulleted impacts of deforestation?

Answer on p. 219

Other disruptions of the drainage basin cycle include the abstraction of water from rivers and lakes for irrigation, industry and domestic consumption, cloud seeding to increase precipitation and building water storage reservoirs.

The hydrological cycle at a local scale

REVISED

This part involves looking at three different aspects of a drainage basin's hydrology.

Water budgets (water balances)

Water budgets show the annual balances between inputs (precipitation) and outputs (evapotranspiration and runoff). They are usually expressed by the formula:

$$P = R + E \pm S$$

where P = precipitation, R = runoff or streamflow, E = evapotranspiration, S = changes in storage.

They therefore can be used to monitor the amount of water held in stores. They are clearly influenced by the type of climate (not just temperatures and precipitation, but also seasonality). Runoff is usually divided into surface flow and base flow. This is an important distinction because in some places there are severe seasonal differences in surface flow, as for example in monsoonal areas. The base flow represents the usually available water.

Since water budgets impact on the availability of water in the soil, they are therefore of great importance to farmers.

> **Exam tip**
>
> You need to be aware of water budget characteristics in tropical, temperate and polar locations.

River regimes

A river regime is the annual variation of the discharge or flow of a river at a particular point along its length. It is particularly affected by climate, geology and soils.

Storm hydrographs

Storm hydrographs show how the discharge of a river varies within a short period of time, such as before, during and immediately after the passing of a storm. They are described as being either 'flashy' (short lag time, high peak and steep rising limb) or 'flat' (long lag time, low peak, gently rising limb).

The key factor is the speed with which the rainfall reaches the river. This depends on:

- a range of different physical features of the drainage basin – size, shape, drainage density, rock type, soil, relief and vegetation
- human factors such as land use and urbanisation. Figure 5.3 shows how increasing urbanisation changes the relative importance of flows. Most notable is the increasing amount of runoff. This, in turn, increases the risk of flooding.

Revision activity

Compare the river regimes of the Amazon (delayed reaction to summer rains), Indus (high flows during the summer monsoon) and Yukon (high flows during input of summer meltwater).

Exam tip

Be sure that you can draw quick sketches of the two different storm hydrographs.

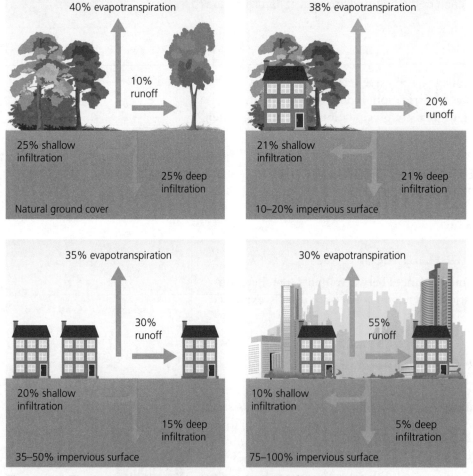

Figure 5.3 **The impact of urbanisation on hydrological processes**

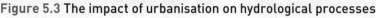

Synoptic theme

Through their control of land use and land-use changes in a considerable number of countries, planners are major players. The decisions they make, for example about the location and amount of urban growth, can clearly affect the behaviour of drainage basin systems. Their decisions will impact most on runoff, water budgets and storm hydrographs.

Now test yourself

4 Why does urbanisation increase runoff and the risk of flooding?

Answer on p. 219

Changes in hydrological systems over time

There are three changes to be understood. The short-term but temporary disruptions of droughts and of floods, and the longer-term disruptions caused by climate change.

> **Exam tip**
>
> At this point in the course, you need to consider the role of planners in managing land use, particularly within urban areas. Why should they do it? How might they do it?

Droughts or water deficits

REVISED

Causes

Droughts are the outcome of deficits within the hydrological cycle. Prolonged shortfalls in precipitation (meteorological droughts) often correlate with ENSO cycles. But droughts of increasing frequency and severity may simply be part and parcel of longer-term changes in climate (see below).

> **Key concept**
>
> The **El Niño–Southern Oscillation (ENSO)** occurs in the Pacific Ocean. But it has a global impact on weather patterns, resulting in more intense storms in some places and drought in others. El Niño events are reversals of the normal directions of ocean currents and winds in the Pacific Basin. Such events usually occur every 7 years and usually last for 18 months.

Now test yourself

TESTED

5 What is the difference between El Niño and La Niña events?

Answer on p. 219

Meteorological droughts give rise to two sequentially related droughts:

- Agricultural drought: the shortfall in precipitation leads to a decline in soil moisture and soil water availability. This has a knock-on effect of reducing plant growth and biomass. Crops yields decline, irrigation systems fail and livestock perish.
- Hydrological drought: reduced precipitation and high rates of evaporation lead to reduced stream flows and falling groundwater levels, as well as reduced storage.

Contributory human factors

The **desertification** of the Sahel (Africa) illustrates how people have helped accelerate the process by a combination of factors:

- overpopulation
- environmental degradation resulting from overgrazing by nomadic tribes and desperate attempts to grow crops
- deforestation as a result of the cutting of fuelwood.

> **Desertification:** The degradation of land in arid and semi-arid areas resulting from various factors, including climatic variations and human activities.

This leads to rural poverty, malnutrition and starvation.

In Australia, droughts vary considerably. While some are intense and short-lived, others last for years; some are very localised, others impact on huge areas of the country for several years, for example the 'Big Dry' of 2006.

Australia illustrates the two sequential droughts: agricultural and hydrological. More than half the farmland has been negatively affected. Water supplies to Australian cities have been threatened as reservoir levels have fallen. The pressure on water resources continues to increase due to the growth of population, with its high water-consuming lifestyles.

> **Exam tip**
>
> Be sure you know the four main human factors contributing to desertification.

Now test yourself

6 Why are Australians characterised as having a high water-consuming lifestyle?

Answer on p. 220

Impacts

Droughts have many impacts, not just on people but on various elements of the physical environment. Droughts have the potential to cause **ecosystem stress** and to test **ecosystem resilience**.

Forests can become stressed by drought, but they usually recover. The same cannot be said for **wetlands**. The term wetland refers to a number of slightly different ecosystems which currently cover about 10 per cent of the Earth's land surface. Wetlands are important for a number of reasons:
- They act as temporary stores within the hydrological cycle.
- They act as giant filters by trapping and recycling nutrients.
- They have very high biological productivity.
- They provide a range of valuable goods and services (see Table 5.2).

> **Ecosystem stress:** Refers to constraints on the development or survival of ecosystems. The constraints can be physical (drought), chemical (pollution) and biological (diseases).
>
> **Ecosystem resilience:** The capacity of an ecosystem to recover from disturbance or to withstand an ongoing pressure, such as drought.

> **Wetlands:** Areas where the soil is frequently or permanently waterlogged by fresh, brackish or salt water. The water may be static or flowing. The vegetation may be marsh, fen or peat.

Now test yourself

7 Why until recently were wetlands drained and infilled? What has changed?

Answer on p. 220

Table 5.2 The value of wetlands

Supporting	Provisioning	Regulating	Cultural
Primary production	Fuelwood and peat	Flood control	Aesthetic value
Nutrient cycling	Fisheries	Groundwater recharge/discharge	Recreational opportunities
Food chains	Tourist attractions	Shoreline protection	Cultural heritage
Life support in carbon cycles		Water purification	

Exam practice answers and quick quizzes at **www.hoddereducation.co.uk/myrevisionnotes**

Despite these valuable functions, and their resilience to drought, wetlands have been drained, dredged and infilled mainly to create farmland or space for urban development. However, after more than 50 years of destruction, the ecological value of wetlands is slowly being realised and some are being given official protection (for example, those designated as a result of the Ramsar Convention).

Floods or water surpluses

A number of physical factors lead to very high flows of water in a drainage basin. Flooding results when the input of precipitation is greater than that which can be carried away by the 'normal' drainage system.

Flood-producing conditions are:
- intense precipitation over a short period (flash flooding)
- sudden snow melt (jökulhlaups)
- unusually heavy and prolonged rainfall (monsoonal rainfall).

Figure 5.4 shows other physical factors involved in flooding.

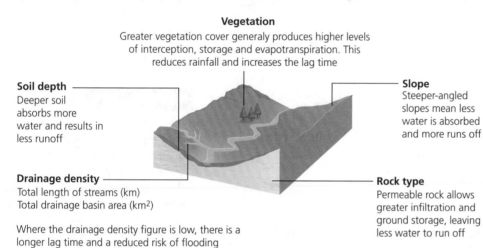

Vegetation
Greater vegetation cover generaly produces higher levels of interception, storage and evapotranspiration. This reduces rainfall and increases the lag time

Soil depth
Deeper soil absorbs more water and results in less runoff

Slope
Steeper-angled slopes mean less water is absorbed and more runs off

Drainage density
Total length of streams (km)
Total drainage basin area (km²)

Where the drainage density figure is low, there is a longer lag time and a reduced risk of flooding

Rock type
Permeable rock allows greater infiltration and ground storage, leaving less water to run off

Figure 5.4 Physical factors contributing to flooding

People can exacerbate the flood risk by:
- changing land use and increasing runoff – for example, by deforestation, overgrazing, draining of wetlands and urban development
- mismanagement of rivers – for example, straightening river channels can improve the flow of water at one location but increase the flood risk downstream
- poor maintenance of rivers and ditches.

The flood events of 2007 and 2012 were reminders of an increasing flood risk in the UK. The summer of 2007 saw more than double the normal amount of rainfall in the three months May–July. At one point in November 2012, the Environment Agency issued nearly 300 flood alerts, mainly in the Southwest and the Midlands. These floods clearly demonstrated the wide range of impacts that this scale of flooding can have:
- death and injury
- damage to the physical infrastructure – roads, railways, water supply and water treatment installations
- the disruption of economic activity – factories flooded, workers unable to get to work
- the inundation of settlements and a huge number of dwellings rendered 'uninhabitable'
- damage to soils and ecosystems.

Impacts of climate change

Inputs and outputs

Most scientists agree that climate change (the outcome of global warming and ENSOs) is enhancing and accelerating the global hydrological cycle.

Changes in temperature, precipitation and evaporation will vary around the world. At present, we are not able to forecast these changes with much certainty or precision. The best we can do is to outline some broader changes. For example, the amount of water in active circulation around the global hydrological cycle is increasing. This means that there is more energy in the atmosphere. This in turn promises heavier precipitation and more severe storms in some climatic regions.

Now test yourself

8 Explain the link between more water circulating around the global hydrological cycle and more energy in the atmosphere.

Answer on p. 220

Stores and flows

Climate change is already affecting stores:
- decreasing the total amount of water held in the form of snow and ice
- deepening the active layer of the permafrost
- lowering water levels in lakes and reservoirs
- reducing wetland storage
- reducing soil moisture.

As for flows, it looks as if more climate extremes will be reflected in an increase in hydrological extremes. There will be more high flows (floods) in some parts of the world and more low flows (droughts) in others. More intense rainfall in some locations will increase runoff rates and reduce infiltration.

> **Typical mistake**
>
> It is not the case that climate change only causes more droughts. This is true for some climatic regions, but there are some regions where the prognosis is more floods.

Future uncertainty

Climate change is the outcome of:
- short-term oscillations (such as ENSO cycles)
- long-term shifts that are part of global warming.

There is uncertainty about how exactly climate and hydrological cycles will change in specific locations, particularly in the long term. This is raising concerns about water management generally. Answers need to be found to key questions:
- How reliable are our projections about drought and flood risks?
- Can we accurately factor in the possibility of more extreme weather events?
- How accurately can we forecast human pressure on water resources?

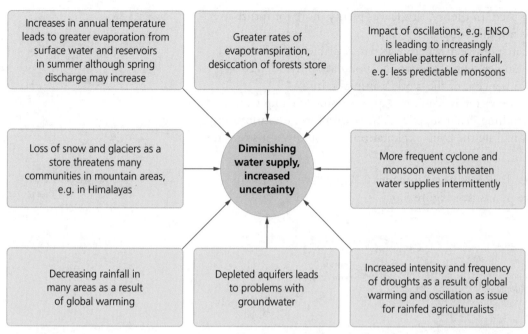

Figure 5.5 The impacts of short-term climate change on water supply

Synoptic theme

The threat that most frightens decision makers and players in the water supply industry is the impact of short-term climate change on water supply (Figure 5.5). Might the possible outcome be a downward path into water insecurity? Is it possible to predict with any accuracy the risks of droughts and floods?

Now test yourself

TESTED

9 Make a list of those factors that create uncertainty about future water supplies.

Answer on p. 220

Water insecurity

The causes of water insecurity

REVISED

Key concept

Water insecurity begins to exist when available water is less than 1,700 cubic metres per person per day. This marks the start of water stress. Below 1,000 cubic metres per person per day, water stress gives way to water scarcity.

Water insecurity: This exists when a population no longer has sustainable access to adequate quantities of water of acceptable quality. Adequate, that is, in sustaining human well-being and socio-economic development and for ensuring protection against water-borne pollution and disease.

A growing mismatch between water supply and demand

It is this growing mismatch between water supply and demand that is creating an increasing amount of **water insecurity** in the world. Remember that the amount of freshwater in the global hydrological cycle is finite. It is estimated that currently 60 per cent of the world's accessible freshwater is being used. In theory, this leaves plenty more for future use. But the situation is not that simple:

● There is a mismatch between where water is available and where the demand is – 66 per cent of the world's population lives in areas receiving only 25 per cent of the world's annual rainfall.
● There is a widening water availability gap as a result of rising water demand and dwindling water supplies. The reason for the diminishing water supplies is mainly the over-exploitation of groundwater stores for irrigation.

These are the main drivers of water insecurity. They have resulted in a number of pressure points (see Figure 5.6).

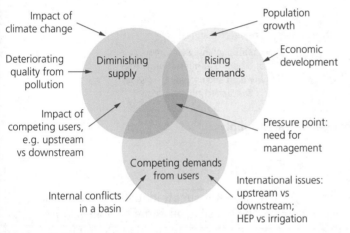

Figure 5.6 Water pressure points

The widening water availability gap means that the world is becoming divided between the 'have-nots' (largely developing countries, particularly in sub-Saharan Africa) and the 'haves' (largely developed countries in temperate latitudes).

Now test yourself

TESTED

10 What are the main drivers of water insecurity?

Answer on p. 220

The physical causes of water insecurity

Water insecurity is the net result of two discrete situations: supply and demand. In the former, the causes are essentially physical, but with some human input:

● Climate: this determines the global distribution of water supply via precipitation. But precipitation varies not only from place to place but also seasonally. Added to this is the impact of global warming which, through the medium of more frequent droughts, is making rainfall receipts less reliable.
● Salt water encroachment: rising sea levels and coastal erosion are leading to the contamination (salinisation) of freshwater by seawater in coastal areas.

- Over-abstraction of water from rivers, lakes and groundwater: the last of these is contributing to salt water encroachment in coastal areas.
- Pollution of water stores and flows resulting from agricultural and industrial effluents.

The human causes of water insecurity

In the case of the demand situation, the causes are largely human:

- Population growth: the global population is now over 7 billion and growing at a rate of 1.1 per cent per annum.
- Rising living standards: increasing the per capita consumption of safe water.
- Industrialisation: 'thirsty' manufacturing and energy generation; pollution source.
- Commercial agriculture: more and more used for irrigation; pollution source.

There is another dimension to water insecurity, namely the distinction between the quantity and quality of available water. Here the key factor is **safe water**. In other words, a country may have access to the required amount of water, but it will be water insecure if only a small proportion of that water is safe. It is predicted that by 2020 30–40 per cent of the world will be suffering from water scarcity. How many governments have made predictions about their national and regional situations?

> **Safe water:** Water that is sufficiently clean to be fit for human consumption and use.

Synoptic theme

A critical task facing water managers is to predict water scarcity. It is necessary to do this at three spatial scales: globally, nationally and regionally. Key players here, apart from the commissioning governments, are meteorologists (forecasting climate change), hydrologists (assessing water stores and stocks), demographers and economists (forecasting demand).

The consequences and risks of water insecurity

REVISED

Water scarcity

An important distinction between physical and economic water scarcity comes into focus here:

- Physical scarcity: the imbalance between water supply and demand, which results in an increasing percentage of available water being consumed. The threshold of physical scarcity is the 75 per cent mark, i.e. more than three-quarters of blue-water supplies are being used.
- Economic scarcity: here the shortfall in available water is related to shortfalls in human resources such as capital, technology and sound governance. The assumption is that the water potential is there, but it waits to be exploited.

Figure 5.7 shows the distribution of both types of water scarcity.

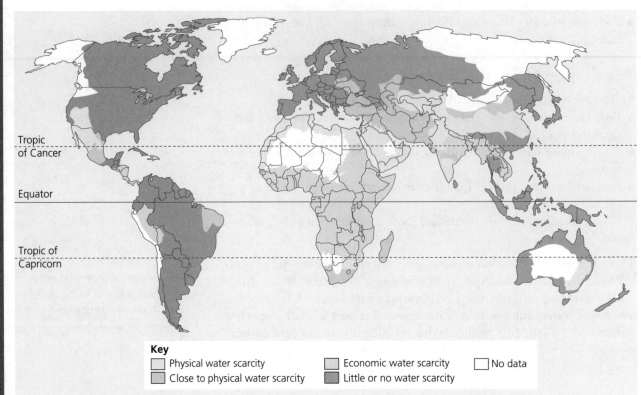

Key

☐ Physical water scarcity	☐ Economic water scarcity	☐ No data
☐ Close to physical water scarcity	☐ Little or no water scarcity	

Figure 5.7 The global distribution of water scarcity

It is easy to think that water supply is somehow free – a gift from nature. Yet in developed and some emerging countries, water is big business. Considerable investments are made in such activities as abstraction, storage, processing and the delivery of water to points of consumption. In short, water comes at a price, and that price is likely to rise with increasing water scarcity. That price also varies spatially.

> **Revision activity**
>
> Make summary notes based on Figure 5.7 about the occurrence of the two types of water scarcity.

Now test yourself

TESTED ☐

11 Why does the price of water vary spatially?

Answer on p. 220

> **Typical mistake**
>
> Do not think that water insecurity is only a problem in drought-threatened developing countries.

The importance of water

The consequences and risks of water insecurity all stem from the basic importance of water to people and their progress. The key uses are:

● human well-being: safe water for consumption, food preparation, sanitation and hygiene
● energy generation
● irrigation
● industry and mining: processing, discharging effluent.

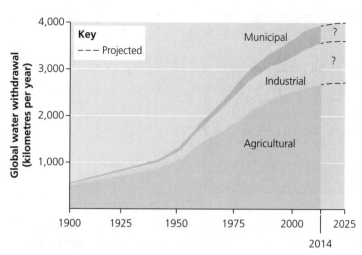

Figure 5.8 Trends in sectoral water use

In Figure 5.8 note the steep rise in the consumption of water by agriculture during the second half of the twentieth century. By comparison, the rise in the use of water by industry has been modest.

Competition and conflict

Competition for water increases as the demand overtakes the available supply. Within a country, competition develops between users. But the users (players) are not equal in terms of their ability. Conflict arises more from the way water is used. For example, use of a river for domestic supply is very likely to be threatened if industry is discharging effluent into the same river.

The drowning of valleys to create storage reservoirs is a common example of water supply becoming a conflict issue.

Competition for water at an international level is more likely to lead to conflict, perhaps even war. Possible scenarios include where a river is shared, such as the Nile by Sudan and Egypt. But here the real upper hand belongs to the owners of the upper reaches of the river (Ethiopia in the case of the Blue Nile and Uganda in the case of the White Nile). In theory Ethiopia and Uganda could abstract as much water as they want and so deprive Sudan of an adequate amount of river water. Sudan could then do the same to Egypt. There is also the potential for conflict where a river forms the border between two countries, as with the part of the Rhine between France and Germany or the Rio Grande between Mexico and the USA.

As yet, there have been no water wars. But it might not be too long before military force is used to protect a country's access to water.

> **Revision activity**
>
> Be sure you have examples of the following:
> - a controversial regional/local reservoir scheme
> - a shared international river (upstream vs downstream)
> - a shared international river (common frontier).

Now test yourself

TESTED

12 Why might it be only a short time before there is an outbreak of water wars?

Answer on p. 220

Different approaches to managing water supply

REVISED

Four broadly different but not discreet approaches to managing water supply are indicated in the specification.

Hard engineering

The classic examples here are the many water transfer schemes that divert water from one drainage basin to another. As such, these usually require constructing huge dams for collecting the water and a large canal to carry that water from an area of surplus to an area of deficit. The most spectacular scheme to date is the South–North Transfer Project in China. Such schemes have been found to have their environmental and social costs (Figure 5.9).

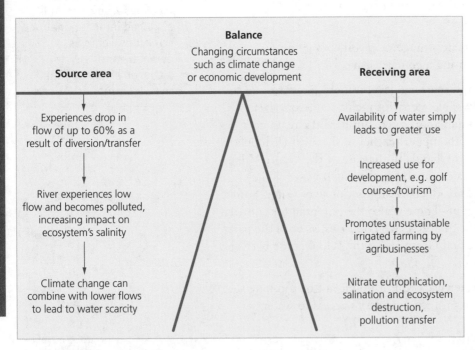

Figure 5.9 Water transfer issues

Desalinisation is a rather different hard-engineering approach which has the advantage of drawing on the seas rather than freshwater supplies. Recent breakthroughs in technology (for example, the reverse osmosis process) have made desalinisation more cost effective, less energy intensive and easier to implement on a large scale.

> **Desalinisation:** The conversion of salt water into freshwater through the partial or complete extraction of dissolved solids.

Sustainable schemes

Key concept

Water sustainability is about ensuring that there are adequate supplies of available water for the benefit of future generations. Water sustainability has three different aspects:

- environmental = freedom from pollution and image, so available as safe water
- economic = ensuring a secure water supply to all users at an affordable price; maximising efficiency of water usage and minimising wastage
- socio-cultural = ensuring equitable distribution of water a) to poor and disadvantaged groups, b) within and between countries.

Revision activity

Name four of the top ten countries with the largest desalination capacities.

This approach is very much focused on water conservation, as well as on the efficiency of water use and the recycling of water. Faced with an acute shortage of water, Singapore has led the way in filtration technology that makes possible the recycling of dirty water.

Synoptic theme

Attitudes to water supply vary very much with climate. Where there is abundant precipitation, there is little concern about levels of water consumption and sustainability. However, where water is scarce and demand puts pressure on supply, very different attitudes prevail: sustainability and efficiency of use becomes priorities.

Now test yourself

TESTED

13 What is the difference between environmental and economic water sustainability?

Answer on p. 220

Integrated drainage basin management (IWBM)

With this approach, river basins are treated holistically. The particular aims are to:

- encourage cooperation between basin users and players
- protect the environmental quality of the river catchment
- ensure water is used with maximum efficiency
- distribute water equitably between its users.

Water sharing treaties

The challenge here is to resolve or neutralise the potential for hostilities over shared water.

Notable advances in achieving the required international cooperation include the Helsinki Rules, which introduced the concepts of 'equitable use' and 'equitable shares', the UNECE Water Convention, which focused on the joint management and conservation of shared freshwater ecosystems in Europe and neighbouring regions, and the EU Water Framework and Directives on issues such as pollution and hydropower.

Exam tip

Be sure you have case study details of i) a catchment shared by a number of countries (e.g. the Nile) and ii) a river forming a national frontier (e.g. the Rio Grande).

Synoptic theme

When dealing with controversial issues, the players indicated here tend to fall into distinct camps: social versus political players; economic versus environmental players.

The world is slowly beginning to appreciate:
- the real value of water as a resource
- the increasing scarcity of water in the face of increasing demand
- the potential of water to become a source of international conflict
- the critical need for international cooperation where there are shared waters.

Revision activity

Note down what you think are the strengths and weaknesses of each of the four different approaches to water supply management.

Skills reminder

You should be familiar with the skills and techniques used in the following investigations of the water cycle and water insecurity:
- analysing flows within hydrological and drainage basin systems
- comparing river regime discharges
- constructing and analysing water budget graphs
- comparing storm hydrographs
- interrogating large databases in terms of trends in floods and droughts
- interpreting synoptic charts relating to drought and flood conditions
- analysing global distributions of water stress and water scarcity
- interpreting indices of water poverty
- identifying seasonal variations in the impact of dams on river discharges.

Exam practice

A-level

* Note that Topics 5 and 6 are examined together in one composite question.

1 (a) Study Figure 1. Explain the difference in soil moisture status between A and D. (3)

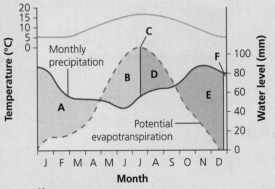

Key
- ☐ Water surplus
- ☐ Soil moisture utilisation
- — Precipitation
- ☐ Soil moisture deficiency
- ☐ Soil moisture recharge
- -- Evapotranspiration

A Precipitation > potential evapotranspiration. Soil water store is full and there is a soil moisture surplus for plant use. Runoff and groundwater recharge.

B Potential evapotranspiration > precipitation. Water store is being used up by plants or lost by evaporation (soil moisture utilisation).

C Soil moisture store is now used up. Any precipitation is likely to be absorbed by the soil rather than produce runoff. River levels fall or rivers dry up completely.

D There is a deficiency of soil water as the store is used up and potential evapotranspiration > precipitation. Plants must adapt to survive, crops must be irrigated.

E Precipitation > potential evapotranspiration. Soil water store starts to fill again (soil moisture recharge).

F Soil water store is full, field capacity has been reached. Additional rainfall will percolate down to the water table and groundwater stores will be recharged.

Figure 1 A water budget graph for southern England, showing soil moisture status

(b) Explain how people increase the flood risk. (6)

(c) Explain the factors increasing the likelihood of water wars. (8)

Answers and quick quiz 5 online

ONLINE

Summary

You should now have an understanding of:

- the nature of human development
- the global hydrological cycle – its stores, fluxes (transfers) and processes
- water in the hydrological cycle available for human use
- physical and human factors affecting drainage basin systems
- water budgets and their impacts
- river regimes and their impacts
- factors affecting storm hydrographs
- the causes and impacts of droughts
- the causes and impacts of floods
- global warming, short-term oscillations and climate change

- the impacts of climate change on trends in precipitation and evaporation
- the impacts of climate change on hydrological stores and flows
- the growing mismatch between water supply and demand
- the causes of water insecurity
- the consequences and risks associated with water insecurity
- the importance of water supply
- the challenges of shared waters
- different approaches to the management of water supply.

A balanced **carbon cycle** is important in maintaining the health of this planet. It operates at a range of spatial and time scales. Physical processes move carbon between stores on land, in the oceans and in the atmosphere. Changes to the most important carbon stores are a result of physical processes. But through the burning of fossil fuels and large-scale deforestation, people are prompting critical changes in those carbon stores. As a consequence, people are contributing to climate change.

The carbon cycle and terrestrial health

Terrestrial carbon stores

REVISED ☐

The carbon cycle

> **Key concepts**
>
> The **carbon cycle** (Figure 6.1) is a closed system: it does not have any external inputs or outputs. So the total amount of carbon is constant and finite. The system has three components:
> - stores: reservoirs where carbon is held
> - fluxes: the flows which move carbon between stores, from one sphere to another
> - processes: the physical mechanisms which drive the fluxes between stores.
>
> **Carbon stores** function as sources (adding carbon to the atmosphere) and sinks (removing carbon from the atmosphere).

The **carbon stores** are:
- the atmosphere: gases such as carbon dioxide and methane
- the hydrosphere (oceans, lakes, etc.): dissolved carbon dioxide
- the lithosphere: carbonates in limestone and fossil fuels
- the biosphere: living and dead organisms.

> **Exam tip**
>
> Be sure you know the four spheres and the form in which they store carbon.

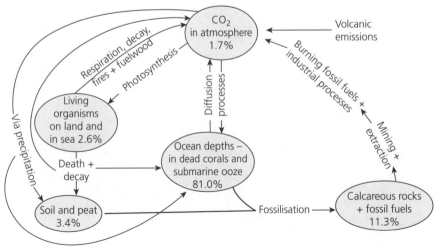

Figure 6.1 The carbon cycle

The amount of carbon dioxide stored in the atmosphere is added to by the burning of fossil fuels and fuelwood and by the respiration of plants and animals. The amount is reduced by precipitation and photosynthesis (Figure 6.1).

Since fluxes are flows, they are measured in terms of their rates. There are significant differences in these rates (Table 6.1).

Table 6.1 A sample of carbon fluxes

Flux	Rate (PgC per year)
Photosynthesis	103
Respiration	50
Volcanic eruption gases	50
Diffusion from ocean	9.3
Vegetation to soil decomposition	5
Weathering and erosion	0.9
Sedimentation/fossilisation	0.2

Geological carbon

Carbon is locked in terrestrial stores as part of a long-term geological cycle. Huge stores of organic matter buried in deep sediments have been turned into fossil fuels (coal, oil and gas). Carbon has also been trapped in sedimentary rocks, such as limestone and chalk, and laid down in seas and large lakes. In these stores the **reservoir turnover** is remarkably long and runs into millions of years.

Geological processes release carbon into the atmosphere by volcanic activity at tectonic plate boundaries and through the chemical weathering of rocks.

> **Typical mistake**
>
> Limestone and chalk are not the only rocks to contain carbon.

> **Reservoir turnover:** The rate at which carbon enters and leaves a store. It is measured by the mass of carbon in any store divided by the exchange fluxes.

> **Now test yourself**
>
> TESTED
>
> 1 Where are the main carbon stores and in what form is the carbon held in each?
>
> Answer on p. 220

Biological processes sequestering carbon

REVISED

Sequestering is the movement of carbon into carbon stores. It has the effect of lowering the amount of carbon dioxide in the atmosphere. The main process responsible for sequestering is **photosynthesis**. But photosynthesis is not confined to land-based plants. Phytoplankton sequesters atmospheric carbon during photosynthesis in surface ocean waters (Figure 6.2).

> **Sequestering:** The long-term storage of carbon dioxide and other forms of carbon.
>
> **Photosynthesis:** The process by which plants capture carbon dioxide from the atmosphere and then store (sequester) it as carbon in their stems and roots. Some will also be stored in the soil.

> **Now test yourself**
>
> TESTED
>
> 2 What is the link between sequestering and photosynthesis?
>
> Answer on p. 220

Oceanic sequestering

Crustaceans (shellfish) depend on carbon extracted from the sea and lakes. But much of the carbon is returned to the sea when they die.

The movement of carbon within the oceans is controlled:

- vertically by **carbon cycle pumps** – there are three of them (Figure 6.2) and between them they deliver carbon dioxide to the sea floor and to the ocean surface for release into the atmosphere
- horizontally by **thermohaline circulation** – this is a global mechanism involving surface and deep ocean currents which are driven by differences in temperature and salinity. So far as the carbon cycle is concerned, warm surface waters are depleted of carbon dioxide by evaporation. But they are enriched again as the conveyor belt circulation drags them along as deep or bottom layers.

> **Carbon cycle pumps:** The processes operating in oceans that circulate and store carbon.
>
> **Thermohaline circulation:** The global system of surface and deep-water currents within the oceans driven by differences in temperature and salinity.

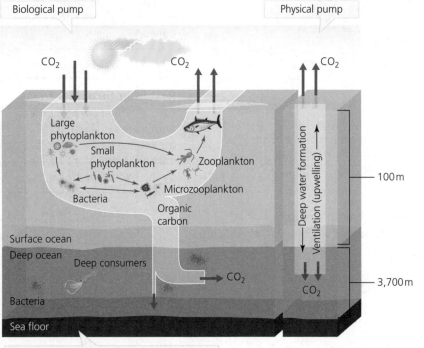

Figure 6.2 Oceanic carbon pumps

Now test yourself

TESTED

3 Name the two factors controlling the movement of carbon within the oceans.

Answer on p. 220

Terrestrial sequestering

Terrestrial primary producers sequester carbon during photosynthesis. Some of this carbon is returned to the atmosphere by respiration from consumer organisms.

Sequestering is a process common to all biomes and ecosystems, but they all differ in terms of their carbon productivity and their carbon storage capacities. Storage is mainly in plants, particularly trees, and soils. The longevity of some trees means that they can serve as stores for tens or

> **Exam tip**
>
> Oceanic sequestering of carbon is important but more difficult to grasp. Be sure you have a general understanding.

hundreds of years. Globally, the most productive biomes are the tropical forests. But carbon fluxes with ecosystems also vary with time: diurnally (most active during the day) and seasonally (most active during the spring and summer).

4 Which biome has the highest carbon productivity and the greatest carbon storage capacity?

Answer on p. 220

Biological carbon

Biological carbon is mainly stored as dead organic matter in soils (20–30 per cent of global carbon). The biological carbon that is not stored is returned to the atmosphere by biological weathering over a period of years. Since all plants are made of carbon, any plant loss to the ground (litter fall) means a transfer or flux of carbon to the soil.

5 Suggest a definition of biological carbon.

Answer on p. 220

Carbon balance and human activities   REVISED ☐

A balanced carbon cycle is important in sustaining other systems of the Earth. Through its control of the amount of CO_2 in the atmosphere, the carbon cycle plays a key role in regulating global temperatures and therefore climate. These changes, in turn, will affect the hydrological cycle. Any increase in the concentration of atmospheric carbon will affect the natural greenhouse effect.

> **Key concept**
>
> In the **greenhouse effect**, the greenhouse gases (carbon dioxide, methane, nitrous oxide) form a layer that is crucial to controlling the temperature of the Earth. As the amount of carbon dioxide in the atmosphere has increased over the last 250 years, so the blanketing effect of greenhouse gases has increased.

Natural greenhouse effect

The concentration of atmospheric carbon (carbon dioxide and methane) strongly influences the natural **greenhouse effect** (Figure 6.3). This concentration determines the global distribution of temperature and precipitation.

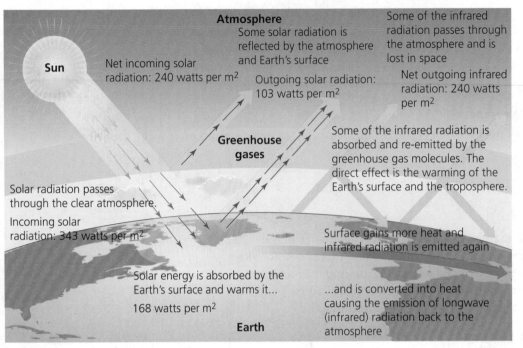

Figure 6.3 The greenhouse effect

The Earth's climate is driven by incoming short-wave radiation. Approximately 31 per cent is reflected back into space by the atmosphere. Nearly half of the remaining 69 per cent is absorbed at the Earth's surface, especially by the oceans, while the other half is re-radiated into space as long-wave radiation. A large proportion of this long-wave radiation is deflected back to the Earth's surface by clouds and greenhouse gases, however. It is this trapping of long-wave radiation that creates the natural greenhouse effect.

Clearly, if the amount of carbon dioxide and methane in the atmosphere increases (for example, as a result of the burning of fossil fuels), then the trapping of more long-wave radiation will raise global temperatures.

> **Exam tip**
>
> It is vital that you have a sound grasp of the greenhouse effect and how it has become the driver of climate change.

Significant regulators of atmospheric composition

Oceanic and terrestrial photosynthesis plays an important role in regulating the composition of the atmosphere. On land, a key factor is soil health, which depends on the amount of carbon stored in it. Carbon helps give the soil its water-retention ability. A healthy soil will, of course, enhance ecosystem productivity. That, in turn, will mean more carbon stored in biomass and more carbon being sequestered from the atmosphere. Of course, partially offsetting this will be respiration of carbon dioxide into the atmosphere.

> **Revision activity**
>
> Draw an annotated diagram showing carbon cycling in soils.

The burning of fossil fuels

Fossil fuels are so called because they were made up to 300 million years ago from the remains of organic material. Once dead, the remains of tiny aquatic animals and plants sank to the bottom of rivers, lakes and seas, were covered with silt and mud, and then started to decay anaerobically. Where the organic matter built up at a faster rate than it decayed, the layers of organic carbon were converted into coal, oil or natural gas by heat and pressure. It is when these fuels are burned that the carbon they have stored for millions of years is released into the atmosphere.

Fossil fuels have been burned at increasing rates since the Industrial Revolution. They continue to be the most important source of primary energy. Without this human intervention, the carbon in fossil fuels would flux very slowly into the atmosphere through volcanic eruptions. The burning of fossil fuels has altered the balance of carbon pathways and greatly speeded up this flux.

It is estimated that about half of the extra emissions of carbon dioxide since circa 1750 have remained in the atmosphere. The rest has been fluxed from the atmosphere into the three carbon stores: the oceans, ecosystems and soils.

The burning of fossil fuels has altered the balance of the carbon cycle, particularly the relative importance of **carbon pathways** and carbon stores. This changing balance is having profound impacts on:

> **Carbon pathways:** The routes taken by flows of carbon between stores.

- global climates: rising global temperatures and therefore rising sea level
- the hydrological cycle: higher temperatures mean increased evaporation and therefore more moisture circulating through the cycle
- ecosystems: changes in their goods, services and biodiversity. Marine ecosystems are threatened by lower oxygen levels, higher rates of ocean acidification and food chain changes.

Now test yourself

TESTED

6 What are the main reasons for burning fossil fuels?

Answer on p. 220

Consequences of the increasing demand for energy

It is the continued burning of fossil fuels to meet the increasing demand for energy that is disturbing the carbon cycle.

Energy security

REVISED

Energy security (Figure 6.4) is a key goal for almost all countries, particularly in this era of growing concern about the use of fossil fuels and the link to climate change.

> **Key concept**
>
> **Energy security** refers to the uninterrupted availability of energy resources at an affordable price. Long-term energy security comes from prudent investments in energy supply and in line with predictions of future demand. In the short term, energy security is about ensuring an uninterrupted energy supply and the ability of the energy system to react promptly to any sudden change in the balance between energy demand and supply.

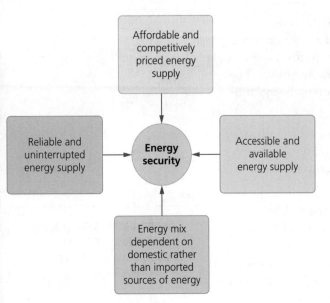

Figure 6.4 Energy security

Rising consumption and energy mix

The rising consumption of energy reflects three factors:
- the growth in the global population
- the rising standards of living being created by economic development
- the essential nature of energy to everyday life.

Now test yourself

TESTED

7 Why do rising living standards lead to higher energy consumption?

Answer on p. 220

Remember that consumption is a function of demand: the greater the demand, the higher the consumption. But there are other factors at work (Figure 6.5).

Figure 6.5 shows the rise in global energy consumption since 1840 and how the **energy mix** has changed.

Energy mix: The combination of different available energy sources used to meet a country's total energy demand.

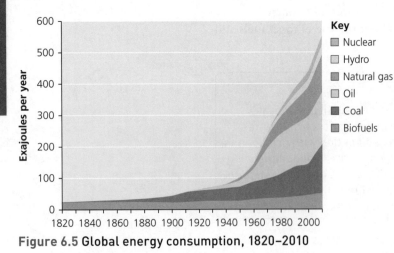

Figure 6.5 Global energy consumption, 1820–2010

Now test yourself

8 Classify the fuels in Figure 6.5 according to the threefold scheme: fossil fuel, recyclable and renewable.

Answer on p. 220

The energy mix has three important dimensions and potential tensions:
● Domestic versus foreign sources: clearly the more the ratio favours the domestic input, the better it is in terms of energy security. However, it might be the case that foreign sources are cheaper.
● **Primary energy** versus **secondary energy**: there might be cost advantages in a reliance on primary energy, but modern economies and societies are highly dependent on electricity.
● Renewable versus non-renewable: given the increasing concern about the burning of fossil fuels and global warming, there is clearly considerable pressure to increase renewable input. However, changing this mix is easier said than done. For example, it may involve covering large areas of both land and sea with wind and solar farms.

> **Primary energy:** The forms of energy found in nature that have not been subjected to any conversion or transformation process, for example water (watermills) and wind power (windmills) or the burning of gas in heating the home and by motor vehicles.
>
> **Secondary energy:** Derived from the conversion of primary energy. The most important secondary energy is electricity, which is derived from many primary sources.

Now test yourself

TESTED ☐

9 Classify each of the following sources of energy as either primary or secondary: geothermal, solar, biofuels and nuclear.

Answer on p. 220

Access to energy resources

Access to and the consumption of energy depend on a range of factors (Figure 6.6).

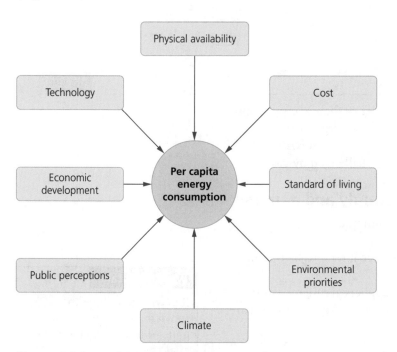

Figure 6.6 Some factors affecting per-capita energy consumption

Energy players

Meeting the demand for energy involves creating **energy pathways** running from the energy producers to the energy consumers. At both ends of such pathways there are influential players (governments, organisations, companies and individuals) with particular involvements in the energy business.

Synoptic theme

Major players and the energy pathways are shown in Figure 6.7.

ENERGY PATHWAY

ENERGY SUPPLY → ENERGY DEMAND

Players	Players	Players
TNCs	TNCs	TNCs
OPEC	Shipping companies	Energy companies
Governments	Pipeline controllers	Governments
		Consumers

Figure 6.7 Players and the energy pathway

Now test yourself

TESTED

10 Name three TNCs involved along the complex length of the energy pathway.

Answer on p. 220

Revision activity

What is OPEC and who belongs to it?

Reliance on fossil fuels

REVISED

Most countries today still derive the greater part of their energy supply from fossil fuels. China and the USA are by far the largest consumers of such fuels. In the twentieth century oil took over from coal as the number one fossil fuel; today oil is now being challenged by gas.

Mismatch between fossil fuel supply and demand

The three fossil fuels need to be considered individually in assessing any mismatch:

● Coal: small mismatch as main producers of coal tend to be the main consumers.
● Oil: a considerable mismatch; the main suppliers are members of OPEC and the consumers are located in Europe. All four BRICs are producers.
● Gas: supply is dominated by the USA and Russia; major importers are Western European countries and Japan.

Typical mistake

It is wrong to think that the move to oil and gas was caused by the exhaustion of coal reserves.

Energy pathways

Energy pathways have been defined above. Again, the three main fossil fuels should be considered separately:

- Coal: there is still a significant global trade. Three of the main coal producers (China, India and the USA) also import coal. Australia and Indonesia are major exporters to Japan, South Korea and Taiwan.
- Oil: the Middle East is the major hub of oil exports.
- Gas: the pathways are not that different from those of oil but there is a major pathway from Russia to Europe.

Unconventional sources of carbon energy

There are four main 'unconventional' sources of fossil fuel:

- Tar sands: Canada is a major exploiter.
- Oil shale: little exploitation as yet.
- Shale gas: exploitation requires controversial **fracking**; the USA is the leading producer and exporter. There is widespread potential.
- Deepwater oil: technological advances in drilling are beginning to make this a viable source. Brazil is leading the way at the moment.

The negative with all these sources is that their use still threatens the carbon cycle. Exploitation methods – pumping, mining and fracking – also pose risks for fragile environments. Yet it is possible that production costs might be lower than those for conventional sources of oil and gas. This is beginning to look true for shale gas. But first there needs to be clear evidence that fracking is safe.

> **Revision activity**
>
> Find out why in 2016 the UK began to import shale gas.

> **Synoptic theme**
>
> Players in the harnessing of unconventional fossil fuels are:
> - exploration companies
> - environmental groups
> - affected communities
> - governments.

Now test yourself

TESTED

11 What might a government's attitude be to the exploitation of unconventional fossil fuels?

Answer on p. 220

Alternatives to fossil fuels

REVISED

Renewables and recyclables

If global carbon emissions are to be reduced, then the world needs to increase its reliance on alternative sources of 'clean' energy. These fall into two categories:

- renewable sources: hydro, wind, solar, geothermal and tidal energy
- recyclable sources: nuclear power and biofuels.

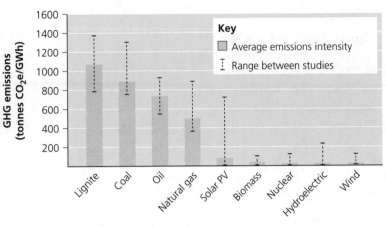

Figure 6.8 Greenhouse gas emissions of the nine electricity-generating fuels

Figure 6.8 confirms how well the renewables and recyclables compare with the four fossil fuels. But note that they are not entirely 'clean' in terms of greenhouse gas emissions.

Now test yourself

TESTED

12 How is it that the renewables and recyclables are all shown in Figure 6.8 as making emissions?

Answer on p. 220

As regards greater use of renewables, there are some negatives:
- Not all countries have physical geographies that contain all or most of these natural resources.
- Increasing the exploitation of these resources is likely to have significant impacts on the environment and on communities.
- People often go off the idea of renewables in the face of a proposal to build a wind or solar farm close to home.
- The construction costs involved in harnessing renewables are high, as are the maintenance costs. So renewable energy is not as cheap as many people like to think.

There are two sources of recyclable energy, nuclear power and biofuels. The risks associated with **nuclear power** are well known. But many developed countries, with their high levels of energy demand, have no other option than to include nuclear power in their energy mix. The UK is one such. The point also needs to be made that the nuclear option, because of its technological and capital demands, is not open to developing countries.

Biofuels

The burning of fuelwood has the longest history. However, there are now a few biofuel crops that are beginning to make their mark on the energy scene:
- Primary biofuels include fuelwood, wood chips and pellets that are used in an unprocessed state for heating, cooking or electricity generation.
- Secondary biofuels are derived from the processing of crops, such as sugar cane, soybeans, maize and cashew. Two types of fuel are extracted (bio-alcohol and biodiesel), both of which are being used as a vehicle fuel and to generate electricity. Brazil is the world leader here.

Typical mistake

It is not the case that renewable sources of energy will completely replace the energy currently derived from fossil fuels. The best hope is to reduce the use of fossil fuels and so lower carbon emissions to 'safe' levels.

Exam tip

Be sure you are familiar with the risks and advantages of using nuclear energy to generate electricity.

Primary biofuels have the disadvantage of encouraging either deforestation or the use of arable land to grow trees. The latter problem also exists with the growing of biofuel crops. Is this a sensible option in an increasingly hungry world? Added to this, there is some uncertainty as to how 'carbon neutral' they are.

Radical technologies

There are two technologies here that would reduce carbon emissions:
- Carbon capture and storage: its potential effectiveness is hindered by the complex technology involved and the doubts about the security of long-term storage.
- Hydrogen fuel cells: this promising technology, making use of an abundant element, is most likely to be used as a source of heat and electricity for buildings and as a power source for electric vehicles.

Links to the global climate system

Human threats to the carbon and water cycles

REVISED

These two cycles are being threatened by human activity, particularly those parts of both that involve the terrestrial biosphere. The biosphere sequesters about one quarter of fossil fuel carbon dioxide emissions. In so doing it slows down the rate of global warming. However, its capacity to do so is being reduced by **land conversion**.

Changes in land use

Land conversion is taking place at an increasing rate in order to meet the growing demand for food, energy and other resources. Any land conversion is almost bound to affect both cycles, specifically reducing both carbon and water stores, as well as the health of soils. Of the range of land conversions, the most serious is undoubtedly deforestation (see below).

Also serious is the conversion of grasslands into cropland, the building of dams and reservoirs, and opencast mining.

Acidification of the oceans

Oceans are important carbon sinks, but due to the increasing uptake of carbon dioxide from the atmosphere, their overall pH is decreasing. This is leading to **ocean acidification**. One of the consequences of this is a serious disturbance of the food web based on coral. Acidification of the oceans is increasing the risk of marine ecosystems reaching a tipping point or threshold of permanent damage. The risk is also being heightened by other stresses such as warming temperatures, more destructive cyclones and pollution. As a consequence, **ecosystem resilience** is being reduced. Crucial here is the speed of the ocean acidification. If it is too fast, oceanic organisms will not have the resilience to adapt and survive (Figure 6.9).

> **Revision activity**
>
> Find out why some people believe that the growing of biofuel crops is not carbon neutral.

> **Revision activity**
>
> Find out what technological challenges, if any, are associated with the development of hydrogen fuel cells.

> **Land conversion:** Clearing a natural ecosystem and using the space it occupied for a different purpose.

> **Exam tip**
>
> Be sure you understand how land conversion affects global warming. Have some specific examples.

> **Ocean acidification:** The decrease in the pH of the oceans caused by the uptake of carbon dioxide from the atmosphere.
>
> **Ecosystem resilience:** The level and speed of disturbance that ecosystems can cope with while still maintaining their original status.

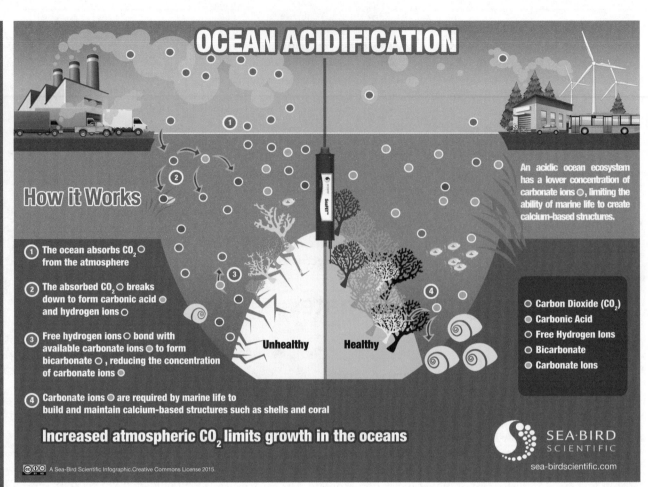

Figure 6.9 Ocean acidification

Climate change

There is now unequivocal evidence that humans and their activities are enhancing the greenhouse effect and that this is leading to climate change. It is estimated that a 2°C rise in global temperatures might lead to a change of climate zone for 5 per cent of the Earth's land area. There is already evidence of an expansion of subtropical deserts and a poleward movement of stormy, wet weather in the mid-latitudes.

Revision activity

Make brief notes on ocean acidification based on Figure 6.9.

Degraded water and carbon cycles and human well-being

REVISED

Deforestation

Of all the possible land conversions none has a greater impact on the carbon and water cycles than deforestation (Figure 6.10). Through their ecosystem services, forests are particularly important to human well-being (Table 6.2). With global deforestation running at 5 million hectares a year and deforestation at its present rate of around 5 million hectares a year, the loss is of fundamental significance to human well-being and survival.

Impact on water cycle
- Reduced intercepted rainfall storage by plants; infiltration to soil and groundwater changes.
- Increased raindrop erosion and surface runoff, with more sediment eroded and transported into rivers.
- Increased local 'downwind' aridity from loss of ecosystem input into water cycle through evapotranspiration.

Impact on carbon cycle
- Reduction in storage in soil and biomass, especially above ground.
- Reduction of CO_2 intake through photosynthesis flux.
- Increased carbon influx to atmosphere by burning and decomposing vegetation.

Figure 6.10 The impacts of deforestation on the carbon and water cycles

> **Exam tip**
>
> You should be sure that you understand the three bulleted impacts on both the water and carbon cycles shown in Figure 6.10.

Table 6.2 Forest ecosystem services and human well-being

Type of ecosystem service		Forest functions and threats
Supporting functions ● Nutrient cycling ● Soil formation ● Primary production	**Provision of goods** ● Food ● Freshwater ● Wood and fibre ● Fuel	● 1.1% of the global economy income ● 13.2 million 'formal' and 41 million 'informal' jobs ● Improve food and nutrition security ● Source of livestock fodder in arid and semi-arid areas ● Fuelwood source for one in three people globally for cooking and boiling drinking water ● A genetic pool: a source for improving plant strains and medicines
	Regulation of Earth systems ● Earth's 'green lungs' regulating climate, floods, disease ● Water purification	Deforestation creates: ● water-related risks (landslides, local floods and droughts) ● increased air and water pollution
	Cultural value ● Aesthetic ● Spiritual ● Educational ● Recreational	● Direct reliance on by many indigenous peoples ● Some cultures and religions see forests as sacred ● Leisure and tourism

For many, the exploitation of environmental resources is a pathway to environmental destruction. However, optimists cling to the **environmental Kuznets curve**, which offers the hope that exploitation will eventually lead to protection.

> **Exam tip**
>
> Make sure you are well informed about the ecosystem services of forests, particularly of the tropical rainforests.

Key concept

The **environmental Kuznets curve** is a hypothetical relationship between environmental quality and economic development (Figure 6.11). Environmental degradation worsens with economic growth until a point is reached when attitudes towards the environment change towards conservation. The critical threshold or tipping point is thought to coincide with a point on the rising curve of income.

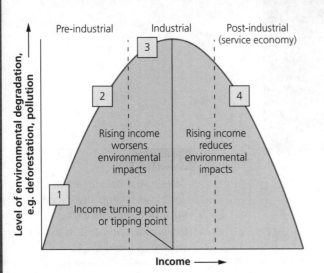

Key
1. UK pre-Industrial Revolution, remote Amazonia today, Indonesia pre-1970s
2. Indonesia today, China in the twentieth century
3. China today
4. UK today

Figure 6.11 The environmental Kuznets curve and tipping points

Now test yourself

TESTED

13 Why might it be said that the environmental Kuznets curve takes a rather optimistic view of the global situation?

Answer on p. 220

Synoptic theme

Global consumers undoubtedly will have different attitudes to the fate of the environment. It is likely that the developing and emerging countries will favour making best use of resources (why should they be prevented from doing so?), while developed countries may be beginning to realise the need to protect and conserve resources.

Higher temperatures and evaporation

Increased temperatures will clearly raise evaporation rates and the quantity of water vapour in the atmosphere. This, in turn, will have a range of impacts on the water cycle, changing:

● precipitation patterns
● river regimes
● water stores (cryosphere and drainage basin).

Exam practice answers and quick quizzes at **www.hoddereducation.co.uk/myrevisionnotes**

These sorts of change will have impacts on agriculture, on hydro power and on the cryosphere. The melting of the Arctic Sea ice is one obvious sign of the last and one that could have profound repercussions.

Ocean health

It is not only acidification that is affecting the oceans; global warming is affecting ocean temperatures and currents, as well as the supply of nutrients and marine food chains. These changes, in turn, are affecting the distribution, abundance, breeding cycles and migrations of marine plants and animals. The well-being of the millions of people who rely directly and indirectly on marine products for food and income is being adversely affected. Hardest hit countries are likely to include Japan and Iceland and many small developing island states, such as the Maldives and St Lucia. It is likely, too, that tourism will suffer.

The risks of further global warming

REVISED

Future uncertainties

There are uncertainties about future global warming and possible contributory factors:

- Uncertainties about natural factors, such as the role of carbon sinks and their capacity to cope with change, possible feedback mechanisms, such as carbon releases from peatlands and permafrost, as well as tipping points relating to forest dieback and the reversal of thermohaline circulation.
- Uncertainties about human factors, such as the future rates of global economic and population growth, the planned reduction in global emissions and the exploitation of renewable energy sources.

> **Synoptic theme**
> The uncertainty of global projections about carbon emissions (futures) and their impact on global systems through climate change.

> **Now test yourself**
> TESTED
>
> 14 Why is there so much uncertainty about future global warming?
>
> Answer on p. 220

Adaptation strategies

Added uncertainty also attaches to whether or not humans can devise **adaptations** that cope successfully with climate change and its diverse repercussions.

Possible adaptation strategies include:

- water conservation and management, e.g. using less water and using it more efficiently
- increasing the resilience of agricultural systems, e.g. developing drought-resistant crops
- land-use planning, e.g. avoiding the development of areas vulnerable to flooding

> **Adaptation:** In the present context, any action that reacts and adjusts to changing climate conditions.

- flood-risk management, e.g. strengthening flood defences
- solar radiation management, e.g. using orbiting satellites to reflect some inward radiation back into space.

Rebalancing the carbon cycle

Can this rebalancing be achieved through **mitigation**? Possible methods include:

- carbon taxation
- switching to renewable energy sources
- increasing energy efficiency
- afforestation
- carbon capture and storage.

But the success of any of these forms of mitigation requires international agreements at a global scale as well as concerted actions at a national level. Experience so far has shown the extreme difficulty of persuading governments to sign up to global-scale agreements (e.g. Kyoto and Paris agreements), while the individual methods listed above have been found to be problematic even when applied at a national level.

> **Mitigation:** In the present context, any actions that either reduce or eliminate the long-term risk and hazards of climate change.

Synoptic theme

Governments, TNCs and people have contrasting attitudes to the challenges presented by the release of stored carbon and the consequent global warming.

Now test yourself

TESTED ☐

15 What is the difference between adaptation and mitigation? Give some examples of each type of action.

Answer on p. 220

Exam tip

Be prepared to give examples of how and why people might hold conflicting views on global warming.

Skills reminder

You should be familiar with the skills and techniques used in the following investigations of the carbon cycle and energy security:

- proportional flow diagrams to show carbon flows and energy pathways
- interpreting maps showing the global distributions of temperature and precipitation
- analysing the energy mixes of different countries by means of graphs
- interpreting maps showing global energy and flows
- comparing carbon emissions from different energy sources
- using GIS to map land-use changes and deforestation
- analysing maps to identify areas at most risk from climate change
- plotting graphs of carbon levels and calculating means and rates of change.

Exam practice

A-level

* Note that Topics 5 and 6 are examined together in one composite question.

1 (a) Study Figure 1. Assess the importance of the oceans in the carbon cycle. (12)

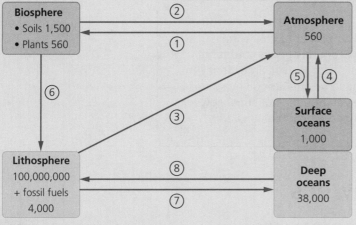

Figure 1 The carbon cycle

(b) Evaluate the merits of mitigation as a strategy to rebalance the carbon cycle. (20 marks)

Answers and quick quiz 6 online

ONLINE

Summary

You should now have an understanding of:

- the global carbon cycle – its stores and fluxes
- geological carbon and its release into the atmosphere
- biological processes sequestering carbon on land and in the oceans
- the importance of a balanced carbon cycle to the health of other Earth systems
- the importance of the natural greenhouse effect
- the impact of burning fossil fuels on climate, ecosystems and the hydrological cycle
- the rising demand for energy
- energy security and the energy mix
- energy players and their roles in securing and supplying energy

- the continuing reliance of economic development on fossil fuels
- the security of energy pathways
- the development of unconventional fossil fuel energy
- exploiting renewable and recyclable energy sources
- the technological challenge of reducing carbon emissions
- human degradation of the carbon and water cycles and the impacts on human well-being
- the uncertainties surrounding future carbon emissions
- responses to global warming – adaptation, mitigation and global agreements.

7 Superpowers

> **Key concept**
>
> **Superpowers** may be distinguished by a number of characteristics, for example by their military strength, economic power and territorial extent. Their pattern of dominance has changed over time. Today the established superpowers are beginning to be challenged by some of the emerging countries. The spheres of influence of superpowers are frequently contested, resulting in geopolitical tensions and even conflicts.

Changing superpowers

The characteristics of superpowers

REVISED

Defining superpower status

A **superpower** is a dominant nation or state with the ability to project and impose its influence outside its borders.

The term 'superpower' dates from the late 1940s when it was used to describe the three dominant world powers at the time: the USA, the USSR and the British Empire.

Four orders of superpower may be recognised. Three are shown in Figure 7.1:
- **hyperpower**
- global superpower
- emerging superpower
- regional superpower.

> **Hyperpower:** An unchallenged superpower that is dominant in all aspects of power (political, economic, cultural, military) – for example, the USA from 1990 to 2010. Is there a hyperpower today?

N = Nuclear weapons

Figure 7.1 The power spectrum of countries by GDP, 2015

Superpower status can depend on five pillars of power. Some nations have all of these pillars, whereas other nations are strong in some types of power but weaker in others:
- Economic power: a large and powerful economy gives nations the wealth to build and maintain a powerful military, exploit natural resources and develop human ones through education.

- Military power: this is used through the threat of military action, or military force can be used to achieve geopolitical goals.
- Political power: this is the ability to influence others through diplomacy or through international organisations.
- Cultural power: this includes how appealing a nation's way of life, values and ideology are to others.
- Resources: these can be in the form of physical resources and human resources.

Typical mistake

Superpower status does not depend entirely on military capability.

Now test yourself

TESTED

1 Which of the five pillars of power do you think is most important? Give your reasons.

Answer on p. 221

Mechanisms for maintaining power

Two forms of power maintenance have existed for centuries (Figure 7.2):

- **Hard power** refers to the way that countries get their own way by using force.
- **Soft power** is the power of persuasion.

Economic power can be thought of as sitting somewhere between hard and soft power. But the most powerful countries utilise smart power (a combination of hard and soft mechanisms) to get their own way.

The use of different types of power is necessary because invasions, war and conflict are very blunt instruments. They often do not go as planned and fail to achieve the aims of those exercising hard power. Soft power alone may not persuade one nation to do as another says, especially if they are culturally and ideologically very different.

Hard power	Economic power	Soft power
←	The spectrum of power	→
• Military action and conquest, or the threat of it • The creation of alliances, both economic and military, to marginalise some nations • The use of economic sanctions to damage a nation's economy	• Economic or development aid from one nation to another • Signing favourable trade agreements to increase economic ties	• The cultural attractiveness of some nations, making it more likely that others will follow their lead • The values and ideology of some nations being seen as appealing • The moral authority of a nation's foreign policy

Figure 7.2 The power spectrum

Changing sources of power

The relative importance of superpower characteristics and mechanisms has changed over time. For example, the German bid for hyperpower status in the first half of the twentieth century was influenced by the ideas of a British geographer, Halford Mackinder, who in 1904 identified a region of Eurasia which he called the **heartland**. This continental land area, protected by invasion from the sea, stretched from Russia to China and from the Himalayas to the Arctic. Mackinder argued that the heartland was the key **geo-strategic location** in the world because controlling it commanded a huge portion of the world's physical and human resources.

Geo-strategic location: A location that commands access to and control over a large territory and its resources.

Now test yourself

2 Explain how superpowers use 'hard' and 'soft' power to maintain their position.

Answer on p. 221

Changing patterns of power

There are different distributions of global power (Figure 7.3):

● A unipolar world is one dominated by one superpower, for example the British Empire.
● A bipolar world is one in which two superpowers, with opposing ideologies, vie for power, for example the USA and the former USSR during the Cold War.
● A multipolar world is more complex, where superpowers and emerging powers compete for power in different regions.

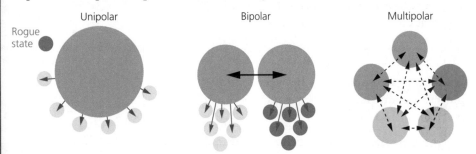

Figure 7.3 The different patterns of power

The colonial era

The high point of superpower polarity was the British Empire during the imperial phases (Table 7.1). Britain managed to rule over a global empire that contained more than 20 per cent of the world's population and accounted for 25 per cent of the world's land area. The Royal Navy dominated the world's oceans during both phases, protecting the colonies and the trade routes between them and Britain.

Table 7.1 The two phases of empire building

Mercantile phase, 1600–1850	Small colonies are conquered on coastal fringes and islands, e.g. New England (now the USA), Jamaica, Accra (Ghana) and Bombay (India), and defended by coastal forts.
	The forts, and navy, protect trade in raw materials (sugar, coffee, tea) and slaves.
	The economic interests of private trading companies such as the Royal African Company, Hudson's Bay Company and East India Company are defended by British armed forces.
Imperial phase, 1850–1945	Coastal colonies extend inland, with the conquest of vast territories.
	Religion, competitive sport (e.g. cricket) and the English language are introduced to colonies.
	Government institutions with British colonial administrators are set up to rule the colonial population.
	Complex trade develops, including the export of UK-manufactured goods to new colonial markets.
	Settlers from Britain set up farms and plantations in colonies.
	Technology, such as railways and telegraph, is used to connect distant parts of the empire.

After the end of the First World War (1918) the distribution of power became increasingly multi-polar. Emerging powers, such as Japan, began threatening the traditional geographical spheres of influence of established superpowers and regional powers. Military power was increasingly important and, as war approached in 1939, an arms race took place, with countries strengthening their naval power.

The post-colonial era

The colonial era came to an end relatively quickly after the end of the Second World War in 1945. Most colonial powers had lost their colonies by 1970. Since then, **neo-colonialism** has emerged in many parts of the developing world.

> **Neo-colonialism:** Refers to an indirect form of control which means that newly independent countries are not masters of their own destinies.

Now test yourself

3 What brought an end to the colonial era?

Answer on p. 221

> **Key concepts**
>
> The **Cold War** was a period of acute tension between the ideologically rival superpowers of capitalist USA and communist USSR. This lasted from 1945 to 1990. During this period, nuclear weapons and the means to deliver them were perfected. The threat of using nuclear weapons added to the tension.
>
> **Hegemony** describes the dominance of a superpower over other countries. Hegemony can be exercised in several ways. The sheer size and multiple capabilities of the USA's military forces give it dominance over world affairs, which deters others from acting against it.

> **Exam tip**
>
> It is important to be able to explain how world power fluctuated between unipolar power, bipolar and multipolar status at different times throughout the twentieth century.

Much of the era from 1945 to 1990 was dominated by the **Cold War**. During this bipolar era the USA became an increasingly global superpower with worldwide military bases aimed at containing the USSR and preventing the spread of communism (Table 7.2). During the same time and since, China has begun to challenge the **hegemony** of the USA.

Table 7.2 The Cold War superpowers compared: USA and USSR

	USA	USSR
Human resources	Population of 287 million in 1989	Population of 291 million in 1991
Physical resources	Self-sufficient in most raw materials; oil importer	Self-sufficient in most raw materials; oil exporter
Political system	Capitalist, free-market economy and global TNCs	Socialist, centrally planned economy; most businesses were state owned
	Democracy with free elections held every four years	Single-party state with no free elections (dictatorship)
Allies	Western Europe through NATO	Eastern Europe (the Warsaw Pact countries) and alliances with Cuba and other developing nations
	Strong economic and military ties to Japan and South Korea	

	USA	USSR
Military power	World's largest navy and most powerful air force, with a 'ring' of bases surrounding the USSR	Very large army, and large but often outdated naval and air force capability
	Large nuclear arsenal and global network of nuclear bases	Nuclear weapons
		Troops stationed in Eastern Europe
	Extensive global intelligence gathering through the CIA	Extensive global intelligence gathering through the KGB
Cultural influence	Film, radio, television and music industry proved a powerful vehicle for conveying a positive view of consumerism, family values, democracy and affluence to a global audience	Exported a 'high' culture message focused on ballet, classical music and art in contrast to the 'popular' culture of the USA
		Strict censorship within the USSR

Geopolitical stability and risk

Different patterns of power bring varying degrees of geopolitical stability and risk (Figure 7.3):

- A **unipolar** world dominated by one **hyperpower** might appear stable, but the hyperpower is unlikely to be able to maintain control everywhere, all the time.
- A **bipolar** world could be stable, as it is divided into two opposing blocs. Stability will depend on diplomatic channels of communication between the blocs remaining open and each superpower having the ability to control countries in its bloc.
- **Multipolar** systems are complex as there are numerous relationships between more or less equally powerful states. The opportunities to misjudge the intentions of others, or fears over alliances creating more powerful blocs, are high and may increase the risk of conflict.

It could be argued that the period between 1910 and 1945 was a multipolar one and that this complex geopolitical situation contributed to two world wars.

Many observers believe that the twenty-first century could be multipolar as countries such as India and China become increasingly powerful while the power of the USA and the EU wanes.

> **Revision activity**
>
> Study Table 7.2 to get the feel of how the USA and the USSR compared as superpowers.

Now test yourself

TESTED

4 Explain the difference between a unipolar, a bipolar and a multipolar world.

Answer on p. 221

Emerging powers

REVISED

Who are the emerging powers?

Future superpowers are likely to emerge from two groups of countries, which overlap:

- The BRIC countries – Brazil, Russia, India and China – were identified as a group of emerging powers in 2001.
- The G20 major economies, a group formed in 1999, is made up of 19 countries plus the EU and includes some potential emerging powers, such as Mexico, Indonesia, South Korea, Saudi Arabia and Turkey.

Strengths and weaknesses

All emerging powers have strengths and weaknesses. For example:

- China has a huge highly educated, technically innovative population, a strong economy and a vast number of military troops. Its weaknesses include an ageing population, serious environmental pollution and a heavy reliance on imported raw materials.
- Russia is a nuclear power with a large military capacity and huge oil and gas reserves. The downside is its ageing and unhealthy population, extreme inequalities and its strained diplomatic relations with the EU and the USA.
- Indonesia has a youthful and potentially dynamic population and large untapped natural resources. Against this there are high levels of urban and rural poverty and internal political instability.
- Turkey has an economy increasingly integrated with that of the EU and has a youthful population with good education levels. However, it has serious internal problems with its Kurdish minority, and is suffering from the political instability in the Middle East.

Now test yourself

TESTED

5 Why is a youthful population thought to be one of the prerequisites of emerging power?

Answer on p. 221

Theoretical explanations

Possible explanations of the changing global pattern of power are to be found in three development theories.

World systems theory views development in a global economic context rather than focusing on individual countries. Three broad categories of economic development categories are recognised:

- core regions – the Organization for Economic Cooperation and Development (OECD) countries and the USA and EU superpowers
- semi-periphery regions – the NICs of Latin America and Asia, including emerging powers such as India and China
- periphery regions – the rest of the developing world.

Dependency theory describes how 'satellite' (periphery) countries provide a range of services to metropolitan (core) countries. The developed countries control the development of developing nations by setting the prices paid for commodities, interfering in economies and using economic and military aid to 'buy' the loyalty of satellite states.

Modernisation theory views the way that countries develop as moving through five stages. It argues that pre-industrial societies develop very slowly until certain preconditions for economic take-off are met. Industrialisation and urbanisation follow and from these processes a country acquires wealth and political power.

> **Exam tip**
>
> Do not ignore the theoretical models that can explain how countries come to gain power. These can form the basis of good examination answers.

Now test yourself

TESTED

6 Explain why and how superpowers can change over time.

Answer on p. 221

The global impacts of superpowers

The global economic system

REVISED

Free trade and capitalism

The global economy is essentially a free-market, capitalist economic system operating throughout most of the world. There are few countries not involved. The promotion of the global economy and **free trade** are in the hands of powerful global **inter-governmental organisations (IGOs)**. The four main players are shown in Table 7.3. The important point here is that the superpowers are able to influence the decisions and policies of those IGOs.

> **Free trade:** The exchange of goods and services free of import/export taxes and tariffs or quotas on trade volume.
>
> **Inter-governmental organisations (IGOs):** Mainly global organisations whose members are nation states that uphold treaties and international law.

Table 7.3 Global organisations and capitalism

World Bank	Makes development loans to developing countries, but within a 'free-market' model that promotes exports, trade, industrialisation and private businesses, which benefits large, developed-world TNCs.
International Monetary Fund (IMF)	Promotes global economic security and stability, and assists countries to reform their economies. Economic reforms often mean more open access to developing economies for TNCs.
World Economic Forum (WEF)	A Swiss non-profit organisation that promotes globalisation and free trade via its annual meeting at Davos, which brings together the global business and political elite.
World Trade Organization (WTO)	IGO established in 1995 that regulates global trade. It has brokered many agreements aimed at promoting open trade and reducing protectionism. Previously known as the General Agreement on Tariffs and Trade.

TNCs and their roles

TNCs come in two distinct forms:
- publicly traded TNCs whose shares are owned by numerous shareholders around the world
- state-owned TNCs that are majority or wholly owned by government.

In many cases state-owned TNCs are large but not well known as their brands are not global.

The dominance of TNCs in the global economy is the outcome of:
- their economies of scale, which mean that they can outcompete smaller companies and, in many cases, take them over
- their bank balances and ability to borrow money to invest in new ventures
- their ability to open up new markets and to expand.

> **Revision activity**
>
> Note the four IGOs in Table 7.3 and how each influences the world economy.

An often overlooked role of TNCs is their promotion of new technology. They invest huge sums in research and development (R&D) to develop new products. Intellectual property law protects these new developments in the form of:

- patents, for new inventions, technologies and systems
- copyright for artistic works, such as music, books and artworks
- trademarks to protect designs, such as logos.

> **Typical mistake**
> Do not underestimate the economic and political clout of TNCs compared with that of superpower governments.

> **Synoptic theme**
> Because of their size, global reach and economic influence, TNCs exercise considerable leverage as players, particularly as far as foreign direct investment is concerned. Critics argue that TNCs have too much power and abuse it in the form of worker exploitation, low wages and environmental pollution.

Global cultural influences

TNCs bring to the countries in which they operate influences from their country of origin. They can have a cultural impact on their global consumers.

The dominance of the USA since 1990, and the economic power of the EU, have led some people to identify the spread of cultural globalisation (commonly referred to as 'Westernisation'). Its characteristics include:

- a culture of consumerism
- a culture of capitalism and the importance attached to attaining wealth
- a white, Anglo-Saxon culture with English as the dominant language
- a culture that 'cherry picks' and adapts selective parts of other world cultures and absorbs them.

Now test yourself

TESTED

7 Explain how TNCs have become dominant economic forces in the global economy.

Answer on p. 221

International decision making

REVISED

Global action

The UN Security Council is the primary global mechanism for maintaining international peace and security. It strives to maintain international law by:

- applying **sanctions** to countries that are deemed to be a security risk, harbouring terrorism, threatening or invading another state or breaching human rights
- authorising the use of military force against a country
- authorising a UN peacekeeping force: troops occupy a country or region under the UN flag to keep the peace in a conflict but do not 'take sides'.

> **Sanctions:** The aim is to force or persuade a country back to the negotiating table without using military force. Sanctions can be diplomatic, economic, military or even sporting.

A striking feature of the past 40 years is the number of times the USA has intervened militarily in foreign countries. It has done this in three ways:

- as part of a UN Security Council action
- together with allied countries as a coalition, but outside a UN remit
- unilaterally, that is, with no support from another country.

Now test yourself

8 Why is there a need for global police?

Answer on p. 221

Alliances

Military alliances are a key element of superpower status. As far as the West is concerned, the most important alliance is the North Atlantic Treaty Organization (NATO).

Other military alliances of note include ANZUS, a treaty involving Australia, New Zealand and the USA. A Russian military alliance, the Collective Security Treaty Organization (CSTO), includes only former USSR republics bordering Russia. The Shanghai Cooperation Organization (SCO) is a strategic partnership involving Russia, China and some of the former USSR republics.

Interdependence between nations is further strengthened by **economic alliances**. Those countries that are members of military alliances are also often involved in economic alliances. For example, many NATO members are EU countries. Figure 7.4 illustrates part of the alliance networks. This creates a powerful axis of economic and military security that reflects the ideology of each bloc. Economic and military alliances overlap with many EU countries that are also NATO members, as well as overlapping between NATO and the North American Free Trade Agreement (NAFTA). NATO, but non-EU, members, such as Iceland and Norway, have a free-trade agreement with the EU (European Free Trade Association, or EFTA).

Revision activity

Find out about three organisations shown in Figure 7.4 – NAFTA, EFTA and TTIP (the Transatlantic Trade and Investment Partnership) – plus ASEAN (not part of the West).

Exam tip

Make sure that you are aware of how the overlap of military and economic alliances supports and enhances superpower status.

Key
- NAFTA
- NATO
- EFTA
- EU
- TTIP (future)

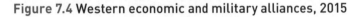

Figure 7.4 Western economic and military alliances, 2015

Geopolitical stability

Figure 7.5 shows the main pillars supporting global security. The system that is in place to help maintain global security was set up in 1945 and revolves around the UN. However, this post-war system is under strain:

- Its leaders – the USA, the UK and France – are not as economically or militarily as powerful as they were.
- There is a strong case for emerging powers – India and Brazil especially – to have more of a say in global security.
- Currently, neither Africa nor Latin America have a say.
- The global financial crisis of 2007–2008 strained the IMF and global financial system to the limit and made many economists wonder whether there was a 'better way'.
- The ongoing threat of global terrorism from al-Qaeda, Islamic State and the Taliban, among others, might suggest that global security cooperation is not all it could be.

Figure 7.5 Pillars of global security

Synoptic theme

The actions and attitudes of IGOs towards maintaining aspects of global security are largely determined by the willingness of members states to act. In most cases this needs strong political will from a group of countries with support from a superpower or emerging power.

Now test yourself

TESTED

9 Which of the four pillars of global security do you think is the most important? Give your reasons.

Answer on p. 221

Global environmental concerns

REVISED

Resource demands

Superpowers have very large resource footprints. Maintaining a large economy, a military machine with global reach and a wealthy population requires energy, mineral, land and water resources.

The high resource consumption of superpowers and emerging powers generates a range of environmental issues:

- Urban air quality is low in emerging power cities due to heavy reliance on fossil fuels.
- The volume of global trade is such that shipping makes a significant contribution to global CO_2 emissions.
- Deforestation and land degradation are issues in some emerging powers as they seek to convert more land into farmland, continue to urbanise, increase the demand for water and increase the use of chemicals in farming to increase yields.

> **Revision activity**
>
> Produce a list showing the ways in which superpower resource demands can cause environmental degradation and contribute to global warming.

Reducing carbon emissions

Recognition of the link between carbon emissions and global warming is increasing. However, there are national differences in the willingness to take action and to sign up to global agreements, such as that forthcoming from the Climate Change Conference in Paris (2015). One aspect of the problem is that developed countries are in a better position to turn to renewable sources of energy. The emerging countries are heavily dependent on fossil fuels and are less willing to sign international agreements.

> **Synoptic theme**
>
> Attitudes and actions over the threat of global warming are influenced by many factors. Generally, Europeans trust the science behind global warming more than Americans do. Economic growth and personal wealth are given a higher priority than environmental issues in some countries, notably the emerging countries.

Now test yourself

TESTED

10 Why might emerging countries be less willing to reduce carbon emissions and sign global agreements on environmental issues?

Answer on p. 221

> **Exam tip**
>
> A range of energy alternatives is now available that can help reduce emissions. But access to those alternatives is far from equal.

Growth in middle-class consumption

A significant future concern related to the growth of BRICS and other emerging powers is that rising affluence will rapidly increase the numbers of **middle-class** consumers. This is obviously positive in terms of development, but it will place a huge strain on resources. This rising middle-class demand, in turn, will affect the availability and cost of key resources, such as rare earth minerals, oil and gas, staple grains and water (Table 7.4).

> **Middle class:** The global middle class can be defined as people with an annual income of over US$10,000. Basically, they are significant consumer spenders.

Table 7.4 Pressure on resources from rising consumption

Food	Water
● Pressure on food supply in emerging powers will result from the nutrition transition and demands for new food types ● Land once used for staple food grains will be converted to produce meat and dairy products ● Without new land, prices could rise, squeezing the poorest	● Some emerging powers already have water supply problems, notably India ● India's situation is likely to be critical by 2030, with 60% of areas facing water scarcity ● Water supply in China, Indonesia and Nigeria could be problematic by 2030, especially in urban areas
Energy	Resources
● Global oil demand was about 95 million barrels per day in 2015 ● By 2030 this is likely to rise, along with coal and gas demand, perhaps by 30% ● Meeting this demand may lead to price rises and/or supply shortages ● Countries with their own domestic supplies (Russia, Brazil) are likely to be in a stronger position than those relying on imports (India)	● Demand for rare earth minerals – used in LCD screens and numerous other hi-tech gadgets – could increase prices ● The demand for lithium-based batteries is very high and could be hard to meet in the future ● Even more basic metals, such as copper, tin and platinum, are at risk of supply shortages and dramatic price changes

Now test yourself

TESTED

11 Explain why the growth of a middle class is thought to be good for development.

Answer on p. 221

Contested spheres of influence

The contest for global influence

REVISED

Tensions over resources

Superpowers are great consumers of resources. Physical resources are important, especially fossil fuels, ores and minerals. Securing access to these resources is vital for both governments and TNCs. Some resources are contested. This could be because:

- the land border between two countries is in dispute
- the ownership of a landmass is in dispute
- the extent of a nation's offshore exclusive economic zone is in dispute or claimed by another nation.

Synoptic theme

Tensions over resources are more likely if countries and/or TNCs have a 'must have' attitude to those resources. If players invest in alternatives (e.g. renewable energy rather than fossil fuels) or give conservation a higher priority than exploitation, then tensions are likely to reduce.

Exam tip

Make sure that you have real-life examples with which you can support your answers in the exam, such as a case study of tensions over the Arctic oil and gas.

Intellectual property

A possible source of tension is over **intellectual property** (IP) rights. A global system of IP has been run since 1967 by the World Intellectual Property Organization (WIPO), a part of the UN. It ensures that TNCs, government agencies, businesses and individuals can protect new inventions, trademarks, artistic works and trade secrets from use by others.

Disregard for international IP treaties and counterfeiting can sour relations between countries, especially between the USA and China:

- TNCs may be reluctant to invest in China, knowing that their profits are likely to be reduced by counterfeiting.
- Lack of action by the Chinese authorities on IP issues might suggest its government is less likely to cooperate on other issues of international law.
- The possibility of trade agreements being made is limited if one side believes the other will not 'play by the rules'.

Political spheres of influence

Political **spheres of influence** can be contested, leading to tensions over territory and physical resources.

A prime example is China's **Island Chain Strategy** in the South and East China seas. This involves claiming sovereignty over small, uninhabited islands as well as artificial islands and so extending its maritime border and economic exclusion zone. So the aims are both military and economic. This strategy is creating tensions with Brunei, Japan, Malaysia, the Philippines, South Korea, Taiwan and Vietnam.

Other current examples include the long-running dispute between India and Pakistan over the ownership of Kashmir, and between Japan and Russia over the Kuril Islands.

> **Intellectual property:** Intangible property that is the result of a person's creativity, such as patents, copyrights, trademarks, etc.

> **Revision activity**
>
> Find out why IP is sometimes criticised.

> **Sphere of influence:** A physical region over which a country believes it has economic, military, cultural or political rights.

> **Revision activity**
>
> Make sure that you have case study examples of both a place where there is tension over territory and physical resources and a place where tension has resulted in open conflict.

Changing relations with developing countries

REVISED

Developing economic ties

Existing superpowers, such as the USA and the EU, have often been accused of having unfair relationships with developing countries. This means relationships based on:

- neo-colonialism: superpowers pulling the economic and political strings of developing countries, despite not ruling them directly, as during the colonial/imperial era
- unfair terms of trade: cheap commodity exports for the developing world (coffee, cocoa, oil, copper) set against expensive manufactured imports from developed countries
- the brain drain of skilled workers from developing countries to boost developed world economies
- local wealthy elites who control imports and exports in developing countries, benefiting from the neo-colonial relationship but having no interest in changing it.

Table 7.5 illustrates how the economic relationship between China and Africa is capable of being interpreted in two contrasting ways: neo-colonialism or development opportunity.

> **Typical mistake**
>
> Do not think that superpowers are magnanimous and generous in their dealings with less powerful countries.

Table 7.5 China's relationship with Africa

Neo-colonial challenge?	Development opportunity?
Infrastructure investments ensure China can export raw materials as cheaply and efficiently as possible	China has invested heavily in roads, railways and ports to export raw materials – infrastructure that can be used by Africans themselves
Skilled and technical jobs are often filled by Chinese migrant workers, estimated to number 200,000 in 2014	Vital jobs are created, especially by large industrial, transport and energy projects, which also modernise the economy
Cheap Chinese imports (clothes, shoes, etc.) have undercut local producers and forced them out of business	Chinese factories and mines bring modern working practices, and technology, to Africa
Much of the FDI brings only temporary construction jobs; there are few long-term jobs in mechanised mines and oil fields	Chinese finance has funded seventeen major HEP projects since 2000, adding 6780 MW of electricity to the continent by 2013
Aid from China is tied to FDI: allow investment and China provides some aid	Investment deals are often accompanied by aid, so the benefits of Chinese money are more widely spread

Synoptic theme

The role of emerging powers in Africa could mirror that of the USA in Taiwan, South Korea and Singapore in the 1960s. The USA's support, both political and economic, helped these states achieve their NIC status. Or could it be that the interest in Africa was simply a repeat of the colonial and imperial model of previous centuries?

Now test yourself

TESTED

12 Give examples of the economic ties between emerging powers and the developing world.

Answer on p. 221

The rising importance of Asia

There is no doubt that Asia is, or very soon will be, the dominant global region. It hosts India, China, Indonesia, Japan, South Korea and Malaysia – all powerful countries, some with emerging and even superpower credentials. It is widely expected that by 2050 Asia will be by far the world's most populous continent and the world's largest by GDP.

The rise of Asia is also creating economic and political tensions within the continent:
- The rivalry between China and India as superpowers will intensify.
- Other countries, such as Japan, Malaysia, Indonesia and the Philippines, are jostling for the third, fourth and fifth places in the regional power rankings.
- Japan feels threatened, particularly by China.
- North Korea poses a security threat, particularly to Japan and South Korea, but even to the USA.

Exam tip

The shifting balance of power can be difficult to predict and although Asian dominance looks set to happen, the experience of Japan is a warning that economic growth may not always continue.

Now test yourself

TESTED

13 Suggest why the rising economic importance of some Asian countries might create political tensions.

Answer on p. 221

Tensions in the Middle East

Of all global regions the Middle East is the one that has proved most troublesome over the past 50 years. It has a number of characteristics that make it a frequent location of tension and conflict. The complexity of Middle Eastern politics, religions, ethnic differences and territorial disputes has led to some intractable and potentially dangerous situations (Table 7.6). In addition, there are the tensions created by **Islamic terrorism**.

Table 7.6 Sources of instability in the Middle East

Religion	Oil and gas	Governance
Most of the region is Muslim, but Sunni (Saudi Arabia, United Arab Emirates) and Shia (Iran) sects are in conflict with each other, both within and between countries	65% of the world's crude oil exports originate in the region; the oil and gas reserves have long been a prize worth fighting over	Most of the countries are relatively new states, at least in their current form; democracy is either weak or non-existent; religious and ethnic allegiances are often stronger than national identity ones
Resources	**Youth**	**History**
Although rich in fossil fuels, the region is short of water and farmland, meaning territorial conflict over resources is more likely	Many countries have young populations with high unemployment and relatively low education levels: the potential for young adults to become disaffected is high	Many international borders in the region are arbitrary; they were drawn on a map by colonial powers and do not reflect the actual geography of religious or cultural groups

Synoptic theme

The differences between Western **ideology** and culture (capitalism, democracy, individual freedom, gender equality and perhaps Christianity) and Islamic ideology and culture (primacy of religion, strict laws and gender discrimination) are hard to reconcile.

Key concept

Islamic terrorism involves many terrorist organisations fighting jihad (holy war) against the non-Islamic West. Their motives are complex and include the religious and ideological belief that war should be fought against all non-Muslims. Many are fighting against what they see as long-term interference by the West in the Middle East. It may also be that recruits to organisations such as IS include young people simply looking for adventure and the thrill of war experience.

Ideology: A system of ideas, beliefs and values that forms the basis of economic or political theory and policy.

Now test yourself

TESTED

14 Why are tensions in the Middle East an ongoing challenge to superpowers and emerging powers?

Answer on p. 221

Economic problems

The USA and the EU are the two largest economic powers, but how long will that remain true? Both face significant economic challenges that may prevent them from maintaining their global importance going forward. The USA is arguably in a stronger position than the EU for two reasons:

- Although consisting of 50 states, which have their own rights and laws, the differences between states are minor and they are not sovereign, unlike the 28 countries that make up the EU. The former are much more likely to agree on policy.
- The USA is not ageing as fast as the EU.

Both superpowers face the ongoing costs of **economic restructuring**. Traditional manufacturing cities have lost jobs and require major investment in regeneration as well as the social costs of coping with rising unemployment.

The economic costs of maintaining superpower status

The USA spends more than US$900 billion annually on maintaining its global supremacy. This includes all military spending and intelligence services, as well as foreign aid and NASA.

China spends around US$200 billion by comparison. Emerging powers have lower labour costs and lower salaries and, to some extent, can increase their military power by copying technologies that were initially developed (at huge cost) by the USA.

> **Economic restructuring:**
> The shift from primary activities and secondary industry towards tertiary and quaternary industry as a result of deindustrialisation. It has large social and economic costs.

> **Exam tip**
>
> The UK may not be considered a superpower but it is one of only a handful of countries to have the technology to launch nuclear warheads almost anywhere in the world.

Now test yourself

TESTED

15 Suggest reasons why some people in the USA might question the cost of maintaining global military power.

Answer on p. 221

The future balance of global power

Figure 7.6 sets out four different superpower futures. Each has different implications:

- US hegemony: would China be prepared to go along with this? China has military power and is keen to extend its maritime limits, but its economy is faltering.
- Regional mosaic: this multi-polar structure is inherently unstable as countries of equal power make complex and competing alliances, with no country acting as the 'global police'.
- New Cold War: China rises to become equal in power to the USA; countries will align themselves with one or the other ideology.
- Asian century: this would cause a fundamental shift in the centre of gravity of the global economy but continue the demand for resources in the West.

Synoptic theme

So uncertain is the geopolitical future that all the governments of superpowers and emerging superpowers, as well as TNCs, undertake 'what if' thinking. The aim is for these key players to have a range of policy options at the ready depending on which 'future' becomes reality. To be prepared is to be forearmed.

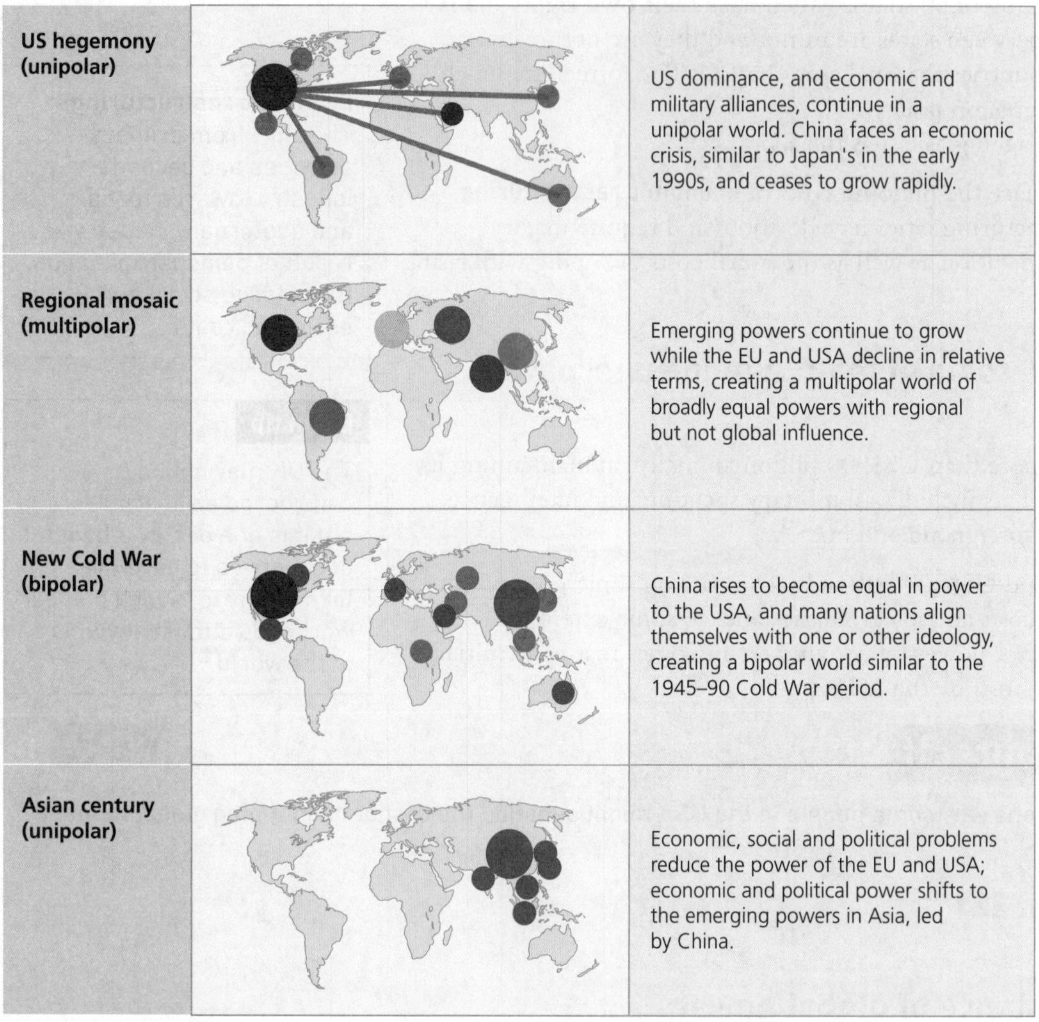

US hegemony (unipolar)		US dominance, and economic and military alliances, continue in a unipolar world. China faces an economic crisis, similar to Japan's in the early 1990s, and ceases to grow rapidly.
Regional mosaic (multipolar)		Emerging powers continue to grow while the EU and USA decline in relative terms, creating a multipolar world of broadly equal powers with regional but not global influence.
New Cold War (bipolar)		China rises to become equal in power to the USA, and many nations align themselves with one or other ideology, creating a bipolar world similar to the 1945–90 Cold War period.
Asian century (unipolar)		Economic, social and political problems reduce the power of the EU and USA; economic and political power shifts to the emerging powers in Asia, led by China.

Figure 7.6 Alternative superpower futures in 2030

Revision activity

Produce a list of the reasons why the future balance of global power is uncertain.

Now test yourself

TESTED ☐

16 Suggest developments that might upset the present balance of power.

Answer on p. 222

Exam practice answers and quick quizzes at www.hoddereducation.co.uk/myrevisionnotes

Skills reminder

You should be familiar with the skills and techniques used throughout the topic of superpowers:

- constructing power indexes using complex data sets, including ranking and scaling
- mapping past, present and future spheres of influence and alliances using world maps
- using graphs of world trade growth using linear and logarithmic scales
- mapping emissions and resource consumption using proportional symbols
- plotting the changing location of the world's economic centre of gravity on world maps
- analysing future gross domestic product (GDP) using data from different sources.

Exam practice

A-level

1 (a) Explain the influence that superpowers have on the global economy. (4)
  (b) Assess the consequences for people and the environment in developing countries as a result of their relationships with superpowers. (12)

Answers and quick quiz 7 online

ONLINE

Summary

You should now have an understanding of:

- how geopolitical power stems from a range of human and physical characteristics of superpowers
- how patterns of power change over time and can be unipolar, bipolar or multipolar
- how emerging powers vary in their influence on people and the physical environment, which can change rapidly over time
- the way in which superpowers have a significant influence over the global economic system
- how superpowers and emerging nations play a key role in international decision making concerning people and the physical environment

- global concerns about the physical environment, which are disproportionately influenced by superpower actions
- how global influence is contested in a number of different economic, environmental and political spheres
- the way in which developing nations have changing relationships with superpowers, with consequences for people and the physical environment
- how existing superpowers face ongoing economic restructuring, which challenges their power.

8 Global development and connections

Option A Health, human rights and intervention

Traditional definitions of development have equated it with economic development. Today, however, while it is still widely acknowledged that economic growth provides much of the driving force behind development, there is a preference for a broader view. This acknowledges that development also has important social, political and environmental dimensions. Development studies now tend to assess progress in terms of **human welfare** and human rights.

The course of global development today is guided by the decisions and geopolitical interventions of national governments and international organisations. These interventions range from development aid to military campaigns and can have a considerable impact on health, well-being and human rights.

> **Human welfare:** The health, happiness, good fortune, prosperity, etc. of a person or a group.

Human development

> **Key concept**
>
> **Human development** has been defined as the process of enlarging people's freedoms and opportunities, and improving their well-being and quality of life. The economic dimension of development is highlighted because it is recognised to be the main driver of the whole human development process. It provides the core of what has been described as the 'development cable'. However, human development is multi-faceted, involving a range of different strands that all contribute to well-being and quality of life (Figure 8.1).

> **Exam tip**
>
> Be sure you are aware of the five different strands of human development shown in Figure 8A.1.

> **Typical mistake**
>
> Do not assume that economic development and human development are one and the same thing.

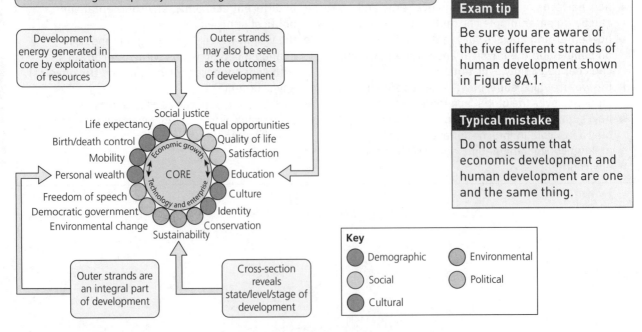

Figure 8.1 The core and strands of the human development cable

Concepts of human development

While most agree that **human development** has an important economic dimension, it is clear that there are some very different perceptions of what it is all about. Those perceptions are coloured by the beliefs, values, morals and codes of conduct of different societies.

A prevailing view today is that human development should focus on:
- health
- life expectancy
- access to education
- human rights.

Measuring human development

Geographers are interested in how the level of development varies spatially – between countries and within countries. The traditional measures used for investigating development in this way have been the economic ones of per capita GDP and per capita GNI. But given the increasing recognition of a broader definition of human development, two relatively new measures are being used:
- Human development index (HDI): based on life expectancy, years of schooling and per capita income (Figure 8.2).
- Happy planet index (HPI): based on sustainable well-being, life expectancy and **ecological footprint** (Figure 8.3).

> **Ecological footprint:** A measurement of the area of land or water required to provide a person (or society) with the energy, food and resources they consume and the waste they produce.

Key

	Development	HDI Index
■	High human development	0.90–1.00
■		0.80–0.89
□	Medium human development	0.70–0.79
□		0.60–0.69
■		0.50–0.59
□	Low human development	0.40–0.49
■		0.30–0.39
■		0.20–0.29
□	not applicable	

Figure 8.2 A view of global development based on the HDI

Key

- ■ All 3 components good
- ■ 2 components good, 1 middling
- □ 1 component good, 2 middling
- □ 3 components middling
- ■ 1 component poor
- ■ At least 2 components poor

Figure 8.3 A view of global development based on the HPI

Now test yourself

TESTED

1 Which do you think provides a better measure of human development, the HDI or the HPI? Give your reasons.

Answer on p. 222

Development goals

While the importance of economic growth as a means of delivering development is widely accepted, it is also being increasingly argued that human development should be targeted at specific non-economic goals, such as improving:

● environmental quality
● life expectancy
● health
● human rights.

Now test yourself

TESTED

2 What are the links between the first three of the four bullet points above?

Answer on p. 222

Education

Education has a pivotal role to play in the development process. It enriches the human resources (human capital) that are vital in the context of economic development. At the same time, it increases awareness of the importance of a whole range of concepts, from human rights to environmental conservation, from healthy living to political freedom.

The educational challenge facing the world is to ensure that:

● everyone has access to a basic education
● access is not conditioned by gender or wealth
● every country is able to offer higher education.

UNESCO is the intergovernmental organisation charged with meeting this global challenge.

Now test yourself

TESTED

3 Why is education so important to a country's well-being?

Answer on p. 222

Human health and life expectancy

The most widely used measure of the general health of a population is **life expectancy**. Data on life expectancy are readily available worldwide. Happily, life expectancies have been rising almost everywhere, but they vary enormously from country to country, for example from 84 years in Japan to 46 years in Sierra Leone.

> **Life expectancy:** The average number of years a person might be expected to live based on the year of their birth.

> **Exam tip**
>
> Remember that there are gender differences in life expectancy. In nearly all countries, female life expectancy is greater. The gender difference is greater in developed countries (it can be as much as five years). It is in only a handful of countries that male life expectancy is greater.

Now test yourself

4 Name two other widely used measures of health at a national level.

Answer on p. 222

Variations within the developing world

These variations in life expectancy and health are explained by differential access to the basic survival needs of food, clean water, sanitation and healthcare. A shortfall in any of these necessities immediately increases the risk of disease, ill health and premature death. Any shortfall tends to impact most on rates of infant and maternal mortality.

Variations within the developed world

Variations here are more to do with differences in lifestyles, levels of poverty and deprivation, and the accessibility and quality of healthcare. It is ironic that the lifestyles of the better-off carry health risks such as obesity, smoking and alcoholism. In countries lacking a national health service, access to healthcare is impeded by poverty.

Variations within countries

Such variations occur within almost all countries and are largely explained in terms of wealth and poverty. In multi-ethnic countries, such as the UK, there are life expectancy differences between ethnic groups. Indigenous people, such as the Aborigines of Australia, often show life expectancies that are up to ten years less than those of the non-indigenous population. A number of factors play a part here, such as the marginalisation of indigenous people, low incomes and lifestyles.

Now test yourself

5 What are the main factors responsible for the variations in human health and life expectancy at the three spatial scales:
- globally: between developing and developed countries
- internationally: between countries falling in either of the above categories
- internally: within countries?

Answer on p. 222

Defining development targets and policies

The relationship between economic and social development

Given that in most countries economic development provides the means that drive and sustain human development, the link between the two is critical. The link is, in effect, in the hands of government. It is government that determines how much of a country's economic wealth should be spent on providing those vital components of human development, such as education, healthcare and other social services. This, in turn, depends very much on governmental attitudes towards **social progress**.

> ### Key concept
>
> **Social progress** derives from the idea that societies have the power to improve their ability to meet basic human needs and to create opportunities for people to improve their lot within society. The pace of improvement is often very slow, but it can be accelerated by:
> * government intervention, for example setting up healthcare, providing housing and free education
> * social enterprise on the part of responsible businesses
> * social activism at a grassroots level that presses for change.

Government attitudes towards social progress are largely conditioned by the type of government or political regime. For example, **democratic governments** are likely to spend more of the national budget on the welfare of the people. They contrast with **totalitarian governments**, which have a reputation for a low level of spending on health and education.

Democratic government: A system of government through elected representatives of the people.

Totalitarian government: A system of government that is centralised and dictatorial. It requires complete subservience to the state, with political control in the hands of elites.

The leading IGOs

Intergovernmental organisations (IGOs) are among the major players in the promotion of global development. Three are particularly influential in the encouragement of economic development: the World Bank, the World Trade Organization (WTO) and the International Monetary Fund (IMF).

The United Nations Educational, Scientific and Cultural Organisation (UNESCO) and the Organization for Economic Cooperation and Development (OECD) are more focused on broader aspects of development, such as health, education and human rights.

Now test yourself

6 Check that you know the specific roles of each of the five IGOs above.

Answer on p. 222

The Millennium Development Goals (MDGs)

Since 2000, global development has been focused on achieving specific targets. There are eight defined MDGs (Table 8.1).

Table 8.1 The Millennium Development Goals

Goal 1: Eradicate extreme poverty	Reduce poverty by half Create productive and decent employment Reduce hunger by half
Goal 2: Achieve universal primary education	Universal primary schooling
Goal 3: Promote gender equality and empower women	Equal girls' enrolment in primary school Women's share of paid employment Women's equal employment in national parliaments
Goal 4: Reduce child mortality	Reduce mortality of under-fives by two-thirds
Goal 5: Improve maternal health	Reduce maternal mortality by three-quarters Access to reproductive healthcare
Goal 6: Combat HIV/AIDS, malaria and other diseases	Halt and begin to reverse the spread of HIV/AIDS Halt and reverse the spread of tuberculosis
Goal 7: Ensure environmental sustainability	Halve proportion of population without improved drinking water Halve proportion of population without sanitation Improve the lives of slum dwellers
Goal 8: Develop a global partnership for development	Use of internet

The MDGs were set for a 2015 deadline. Since then a 2030 Agenda for Sustainable Development has been agreed by world leaders. The 17 goals of **sustainable development** go much further than the MDGs and are more focused on the root causes of poverty and the universal need for a style of development that works for all people and is more aware of the environment (Figure 8.4).

Sustainable development: Development that meets the economic, social and environmental needs of today's population without compromising the ability of future generations to meet their own needs.

Figure 8.4 Sustainable development goals for 2030

Now test yourself   TESTED ☐

7 What is meant by sustainable development?

Answer on p. 222

Revision activity

Make some brief notes on how the SDGs differ from the MDGs.

Human rights

> **Key concept**
>
> **Human rights** are the moral principles that underlie standards of human behavior, the inalienable rights to which a person is entitled regardless of their nationality, language, religion, ethnicity or gender. The rights include liberty, freedom of movement and speech, personal security and access to education and justice.

The importance of human rights

REVISED ☐

As far as the UK is concerned, there have been three landmark statements setting out **human rights**:

- Universal Declaration of Human Rights (UDHR) (1945): not really binding and therefore unenforceable at an international level.
- European Convention on Human Rights (ECHR) (1953): this and the UDHR were a response to human rights violations that had occurred during the Second World War (1939–1945).
- UK Human Rights Act (1998): recognises the human rights set out in the ECHR and comprises 14 articles.

Also be aware of the Geneva Convention (1949), which set out rules that apply only in times of armed conflict.

> **Revision activity**
>
> You should be aware of the range of human rights and the background to the three important statements listed here.

Differences between countries

REVISED ☐

Human rights versus economic development

If pressed, most governments would give the promotion of economic development a higher priority than that of human rights. But that is not to suggest a disregard for human rights. The difficulty here is that there are few, if any, widely recognised measures relating to the status of human rights, measures that might be used to compare countries. Figure 8.5 is a map produced by Freedom House. Countries are rated in terms of political freedom and respect for civil liberties. Each country is broadly classified as free, partly free or not free.

> **Revision activity**
>
> Make summary notes about the 'free' and 'not free' distributions shown in Figure 8.5.
>
> Can you detect any possible correlations between these distributions and types of political regime?

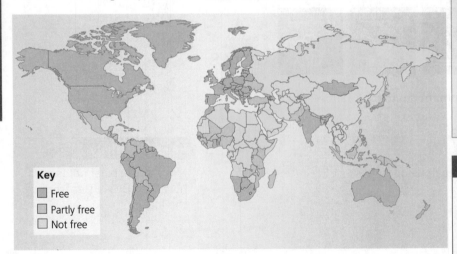

Key
- ☐ Free
- ☐ Partly free
- ☐ Not free

Figure 8.5 Freedom in the world, 2015

> **Exam tip**
>
> To illustrate contrasting attitudes to economic development and human rights, compare North and South Korea.

The transition to democracy

The question arises: is there any correlation between economic development and human rights? Of the ten top economic superpowers, six are long-established democracies with a satisfactory current human rights record: the USA, Japan, Germany, France, the UK and Italy. As for the other four, China and Russia have poor records, while the situation is rather better in the fledgling democracies of Brazil and India.

Political corruption

The term **political corruption** immediately suggests election rigging, but it can take other forms, such as:

- allowing private or business interests to dictate government policy
- taking decisions at a government level that benefit those who are 'funding' the politicians
- diverting foreign aid and scarce resources into the private pockets of politicians.

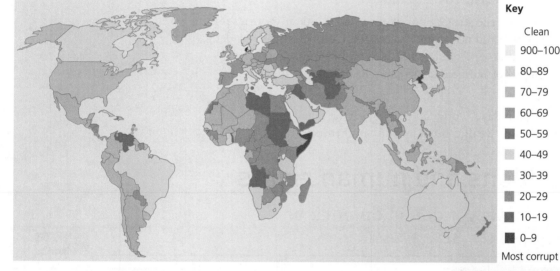

Key

Clean

	900–100
	80–89
	70–79
	60–69
	50–59
	40–49
	30–39
	20–29
	10–19
	0–9

Most corrupt

Figure 8.6 A global view of corruption based on the corruption perception index, 2014

A comparison of Figures 8.5 and 8.6 suggests a partial correlation between 'not free' government and high levels of corruption.

> **Revision activity**
>
> For examples of corrupt regimes, you might look at the cases of Myanmar or Zimbabwe.

Differences within countries

REVISED

Discrimination

The most widespread discrimination within countries is that based on gender and ethnicity. It was only in the twentieth century that women in most parts of the developed world gained equal rights and opportunities. Some would argue that even in these countries the employment playing field is not an even one when it comes to wages, salaries and promotion. However, there remain many parts of the world where women continue to be regarded as second-class citizens.

Although discrimination on the basis of ethnicity has been legislated against in many countries, it may still occur in an insidious manner, as for example in the housing market, at the workplace and in terms of access to higher education.

Differences in health and education

There is plenty of evidence to suggest a broad correlation between human rights on the one hand and access to health and education on the other. It has been decreed by the UDHR that such access is one of the most basic of human rights. However, it is clear that the failure to respect the human rights of indigenous people has impeded their access to education and healthcare.

However, it is likely that the correlation also works in the other direction, namely that education might be expected to lead to a greater respect for human rights.

The growing demand for equality

Within many countries, the struggles of women and ethnic minorities for equality have been long and not always successful. Compare the struggles of women in Afghanistan or Bolivia with those in European countries. Compare the struggles of indigenous people within Brazil with those in Australia.

A possible verdict on today's world: all people may be equal, but they are not so when it comes to wealth, freedom and opportunity.

Interventions and human rights

Geopolitical intervention in defence of human rights

REVISED

Range and motives of intervention

It needs to be understood that **geopolitical interventions** are made for reasons other than just the defence of human rights. Other possible motives include:

- extending development aid to the poorest and least developed countries
- encouraging education and healthcare
- strengthening security and political stability
- promoting and protecting trade
- accessing resources
- offering low-interest loans
- increasing global or regional influence.

> **Geopolitical intervention:** Occurs when and where a country exercises its power in order to influence the course of events outside its borders.

From these motives, it can be seen that there are three main intervention mechanisms or pathways:

- development aid (see below)
- economic support: for example, trade, investment
- military support: for example, training and equipping the armed forces of another country, helping to deal with insurgents and terrorists, all-out military occupation.

The specification focuses on the first and last of these inventions (see below).

International intervention players

The main players are:

- individual governments, often those of the superpowers
- IGOs such as the UN, the EU, the World Bank and the WTO
- NGOs such as Human Rights Watch, Amnesty International, Oxfam and MSF (Table 8.2).

Table 8.2 Some major NGOs in human development

Organisation	Founded	Mission
Amnesty International	1961	Founded in the UK and focused on the investigation and exposure of human rights abuses around the world. Takes on both governments and powerful bodies, such as major companies. Today it combines its considerable international reputation with the voices of grassroots activists on the spot to ensure that the UDHR is fully implemented. It also provides education and training so that people are made aware of their rights.
Human Rights Watch	1978	Founded under the name of Helsinki Watch to monitor the former Soviet Union's compliance with the Helsinki Accord (aimed at reducing Cold War tensions). Like Amnesty International it is constantly on the lookout for violations of the UDHR. It is not frightened to name and shame non-compliant governments through media coverage and direct exchanges with policymakers.
Oxfam	1942	Founded in the UK to help deal with the hunger and starvation that prevailed during the Second World War. Today it has three main targets: development work aimed at lifting people out of poverty and improving health (safe water and sanitation); assisting those affected by conflicts and natural disasters; and campaigning on a range of issues, from women's rights to the resolution of conflicts.
Médicins Sans Frontières (Doctors Without Borders, MSF)	1971	Founded in France with the belief that all people have the right to medical care regardless of race, religion or political persuasion. Today it provides healthcare and medical training in about 70 countries and has a reputation for providing emergency aid in conflict zones. It remains independent of any economic, political or religious influences.

Conditional intervention

Some Western governments have sought to combat violations of human rights elsewhere in the world. They have done so through various offers of 'help' in exchange for undertakings to respect human rights. The 'help' has been mainly in the form of:

- humanitarian aid
- preferential trade deals
- investment
- military protection.

However, such interventions can easily be interpreted by other governments as threatening the sovereignty of the countries receiving such 'help'. A primary example would be the intervention by the Russians in the Ukraine undertaken because of the need the Russians felt to protect the human rights of the ethnic Russians living there.

But there are other kinds of condition, for example the development aid offered by China to African countries in exchange for preferred access to resources that it needs.

Development aid

> **Key concept**
>
> **Development aid** is aimed at development in the broader sense, such as safeguarding human rights and improving human welfare and quality of life. Nonetheless, in many instances it also has an economic dimension.

The range of development aid

Development aid can vary in scale, from installing a village well to constructing a large irrigation project; financially, from a small charitable gift to a global appeal raising millions of pounds; in time scale, from short term (for example, emergency aid) to long term (for example, disease eradication programmes); and in the mix of providers, from local charities to major IGO and NGO players.

Development aid has three main delivery routes:
● bilaterally: directly from one country to another
● multilaterally: indirectly through donations by individual governments to the major IGO players
● charitably: through individual donations to NGOs and their emergency appeals.

Basically the global flow of development aid is from developed to developing countries. Table 8.3 shows the main donors in terms of **official development assistance (ODA)**.

A growing concern in some countries is the size of their aid budgets. Should they be donating more or less?

> **Official development assistance (ODA):** A measure used by the OECD as an indicator of the flows of international aid. Flows are transfers of resources, either in cash or in the form of commodities and services.

Table 8.3 The top ten ODA donors, 2013

ODA (US$ billion)		ODA (% GNI)	
USA	31.55	Norway	1.07
UK	17.88	Sweden	1.02
Germany	14.06	Luxembourg	1.00
Japan	11.79	Denmark	0.85
France	11.38	UK	0.72
Sweden	5.83	Netherlands	0.67
Norway	5.58	Finland	0.55
Netherlands	5.44	Switzerland	0.47
Canada	4.91	Belgium	0.45
Australia	4.86	Ireland	0.45

Now test yourself

TESTED

8 Which do you think is the better indicator of a country's generosity: total ODA or ODA as a percentage of GNI? Give your reasons.

Answer on p. 222

Positive impacts

Development aid has been fairly successful in a number of different contexts:

- in dealing with life-threatening and often contagious diseases, eliminating some (for example, polio) or limiting others (for example, malaria and Ebola)
- in fighting extreme poverty
- in improving access to education and healthcare
- in providing safe water and sanitation
- in stimulating the creation of jobs.

Negative impacts and concerns

Development aid is coming in for criticism on a number of different counts:

- Aid in the form of capital grants and loans is now thought to be inappropriate, if only because it tends to increase the receiving country's indebtedness. Technical assistance and skills training are now preferred.
- There are concerns about the recipients of development aid. For example, large amounts of UK aid are still going to India, which is an emerging country on the up. Are there not countries more deserving of aid?
- A worrying amount of development aid is being siphoned off by corrupt officials and also used as bribes (for example, **land grabbing** in Kenya).
- It encourages the growth of elites in receiving countries who disregard the human rights of others.
- It is thought to encourage dependency rather than stimulate enterprise and self-advancement.
- The involvement of TNCs is questioned because of **profit leakage** and the considerable environmental damage they cause in the exploitation of resources (for example, the exploitation of oil in the Niger Delta).

> **Land grabbing:** The acquisition of large areas of land in developing countries by domestic and transnational companies, governments and individuals. In many instances, the land is simply taken over and not paid for.
>
> **Profit leakage:** The process whereby the profits made by a business are not retained and reinvested in the country where they were made.

> **Exam tip**
>
> Be sure that you are able to cite at least three of these concerns about development aid.

Military interventions

REVISED

Defending human rights

Protecting human rights has often been used as an excuse for military intervention. But whether or not that excuse was valid is another matter. All too often, there have been ulterior motives.

Providing military aid

Military aid can take various forms:

- providing military equipment and armaments
- training the military forces of the receiving country
- providing military personnel: troops, advisors.

A contentious aspect of selling arms to another country is that the supplying country often has little control over how those arms are used and against whom. The UK's supply of military equipment to Saudi Arabia has recently been questioned as some of that equipment has been used in attacks on Yemen.

Waging war on terror and torture

The world today is troubled by Islamist terrorist organisations such as the Taliban, al-Qaeda and Daesh (incorrectly known as IS). The last is causing much trouble in the Middle East, particularly Iraq and Syria, and has mounted occasional attacks in other parts of the world (Figure 8.7). As a result, the Western superpowers find themselves increasingly embroiled in a war on terror.

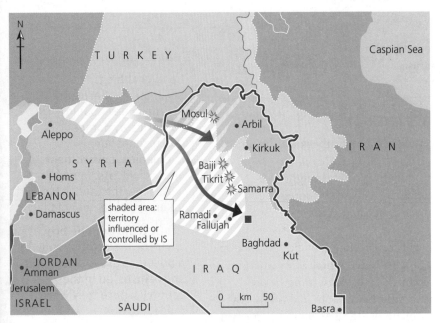

Figure 8.7 The spread of IS in the Middle East

The West's feeling that it should intervene militarily against IS is motivated by three main concerns:

● the political stability of the whole of the Middle East, and possibly that of North Africa
● protecting access to the oil reserves in both regions
● the atrocious abuse of human rights.

Now test yourself

TESTED

9 What are the three motives for taking action against IS?

Answer on p. 222

The military intervention needed here is more one of covert surveillance and intelligence gathering, as well as providing Arab troops on the ground with aerial cover, equipment and training.

The issue of **rendition** and torture as a means of obtaining vital information is a challenging one:

> **Rendition:** The practice of sending foreign criminal or terrorist suspects covertly to be interrogated in a country where there is less concern about the humane treatment of terrorist suspects.

- Whose rights are more important, those of the terrorist not to be tortured or the right to life of those innocent people who could become the victims of terrorism?
- Should there be limits to the actions a government undertakes to ensure national security and territorial integrity?

Revision activity

What does the UN Convention against Torture (1987) stipulate about the treatment of suspects?

Now test yourself

TESTED

10 What is the link between rendition and torture?

Answer on p. 222

The outcomes of geopolitical interventions

Evaluating interventions

REVISED

Measures

Given the diversity of interventions (military, development aid, etc.) and hoped-for outcomes, there are many possible ways of assessing whether or not a particular intervention has had a successful outcome. Table 8.4 shows some possible measures for evaluating interventions aimed at human development and human rights.

Among these measures, there is no single 'silver bullet'. However, there is much to be said for using more than one measure. It is also the case that progress in human development is easier to measure than progress in human rights.

Table 8.4 Possible measures for evaluating intervention outcomes

Intervention target	Possible measure
Human development	Life expectancy Provision of healthcare (doctors per 100,000) Literacy rate (% of population) Quality of physical infrastructure (% with access to safe water and sanitation) Per capita GDP or GNI
Human rights	Freedom of speech Gender equality (gender index) Democratic elections Respect for minorities Recognition of refugee status

Human rights indicators

Potential indicators of progress in human rights tend to be qualitative rather than quantitative. Of the five measures shown in Table 8.4, possibly the incidence of democratic elections is the most telling. The fact is that broad respect for human rights is more likely to flourish in a democracy than in a one-party state without any form of opposition.

Now test yourself

TESTED

11   Explain why it is more difficult to measure progress in human rights.

Answer on p. 222

Importance of economic growth

It is accepted that economic growth provides the fuel needed to drive human development. But equally it is recognised that a serious tension often exists between economic growth and human rights. This is particularly the case if a country is keen to fast-track that growth. This is well exemplified by today's China. Even so, many countries today prefer to have their development status assessed by economic measures such as per capita GDP and per capita GNI.

Evaluating development aid

REVISED

Development, health and human rights

Given that much development aid is targeted at improving health and human rights, this should make evaluation rather more straightforward. What becomes clear is a mixed record, with successes in countries such as Botswana and Guyana contrasting with serious shortcomings in, for example, Haiti and Ivory Coast. Corruption is one factor helping to explain the lack of achievement.

> **Typical mistake**
>
> Not all development aid is provided for humanitarian reasons.

Impact on inequalities

Recipients of development aid also differ in terms of the impact of that aid on internal economic inequalities. In some countries, such inequalities have been reduced (for example, in Cambodia and Senegal) while in others, such as Bangladesh and Honduras, they have increased (with corruption again being a factor). But any increase in inequalities almost inevitably hits the poor and least powerful sectors of society. For them, there are negative impacts on health, life expectancy and human rights.

Now test yourself

TESTED

12   What statistical measure is widely used in the assessment of spatial inequalities?

Answer on p. 222

Superpower objectives

There is no doubt that the superpowers have become donors of development aid for a range of motives that are not necessarily humanitarian. They may be motivated by the need to secure:

- a presence in geo-strategic locations
- future supplies of resources
- military alliances with other countries
- a global sphere of influence.

Now test yourself

TESTED

13 Which are today's leading superpowers?

Answer on p. 222

Evaluating military interventions

REVISED

The most obvious evaluation criterion is whether or not the intervention has been successful – has it achieved its objectives? Whichever way the verdict goes, there will be costs.

Costs

The costs of military intervention fall into two broad categories:
- military: for example, troops killed and wounded; ammunition spent; equipment lost
- civilian: for example, number of deaths, injured and displaced (refugees); damage to, and destruction of, homes, public buildings and infrastructure; disruption of business, etc.

The above are essentially short-term costs. There will also be longer-term costs in the area of intervention depending in part on whether or not the intervention has been successful, such as:
- the time taken to fully recover and get back to normal after the cessation of hostilities
- the impacts of a loss of territory or sovereignty
- a possible loss of human rights.

> **Typical mistake**
>
> Do not think of military intervention as just the movement of troops into a conflict zone.

Costs of non-military intervention

What the specification has in mind here is peacekeeping interventions by the UN, as for example currently in Ivory Coast. It is a possible way of

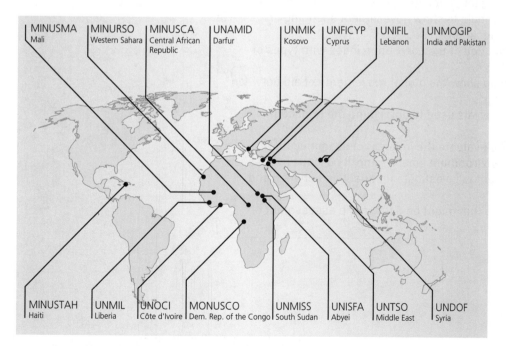

Figure 8.8 The locations of UN peacekeeping operations, 2015

navigating away from a conflict situation. UN peacekeeping is guided by three basic principles:

- all parties in a conflict must consent to the intervention
- the need to show impartiality
- non-use of force except in self-defence and defence of the mandate.

As of 2015, the UN was involved in 16 different peacekeeping operations (Figure 8.8). The record of past operations has not been entirely successful, particularly when it has come to putting in place the right sort of governance.

Costs of no action

Faced with threats to human well-being and human rights, the global community has three options:

- turn a blind eye and do nothing – this may well mean that the threats worsen
- make a limited intervention (military) to deal with the short-term threat – this may bring short-term relief but longer-term costs
- make an extended intervention – but not necessarily a military one.

Revision activity

Make notes about the distribution of the peacekeeping operations shown in Figure 8.8.

Exam tip

It would be worthwhile to have some details of one of the 16 peacekeeping operations.

Now test yourself

TESTED

14 Why might it be better not to make an extended military intervention?

Answer on p. 222

Skills reminder

You should be familiar with the skills and techniques used in the following investigations of health human rights and interventions:

- comparing different measures of development using ranked data
- analysing the relationship between health and life expectancy using scattergraphs and correlation techniques
- using proportional symbols to show government spending on health, welfare and education
- correlating the distribution of the corruption index with types of government
- using flow-line maps to show the global movements of aid between donors and recipients
- evaluating source materials used to determine the impact of development aid
- interpreting images to evaluate the impacts of economic development on the environment where minority groups live
- using the Gini coefficient to investigate inequalities between and within countries
- critical analysis of data that might be used in the assessment of interventions.

Exam practice

A-level

1 (a) Study Figure 1.

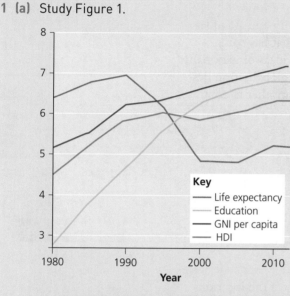

Figure 1 Botswana's human development index and its three components since 1980

(i) By how much did the HDI increase since 1980? (1)

(ii) Suggest **one** reason for the change in life expectancy in Botswana. (3)

(b) Suggest reasons why access to education is such an important human right. (6)

(c) Explain why indigenous peoples are often the target of discrimination. (8)

(d) Evaluate the view that development aid is the best intervention that the developed world can make in the developing world. (20)

Answers and quick quiz 8A online

ONLINE

Summary

You should now have an understanding of:
- the nature of human development
- the importance of economic growth to the delivery of human development
- the importance of education
- the links between health and life expectancy
- variations in health and life expectancy, both between and within countries
- the major players in global development
- setting development goals
- human rights and international agreements
- how countries vary in both their definition and protection of human rights
- significant variations in human rights within countries, which are reflected in different levels of social development

- discrimination based on gender and ethnicity, which remains widespread
- geopolitical intervention in defence of human rights, which can take several different forms
- how development aid takes many different forms, but it does have its critics and the record is a mixed one
- how development aid of an economic kind can have serious negative impacts on the environment and culture
- how military interventions are frequently justified in terms of human rights
- evaluating the outcomes of different types of geopolitical intervention.

Option B Migration, identity and sovereignty

Globalisation means increasing movements of capital, goods and people. Migration flows, both internal and international, are now running at record levels. International migration not only alters the ethnic composition of populations, it also changes national identity and attitudes to it. Indeed, the growing interdependence of countries that has come with globalisation is creating tensions, most notably with ideas of national identity and sovereignty. The fact that the governance of globalisation lies in the hands of a small number of global organisations is another factor encouraging the rise of nationalism in some countries.

Impacts of globalisation on international migration

Internal and international migration

REVISED

A new pattern of labour demand

Globalisation has generated much new employment and it is the locations of this employment that are determining the directions and destinations of **economic migrant** flows. This new pattern makes sense if considered in terms of the **core–periphery model**.

National core–periphery systems have been strengthened not only by rural–urban migration but also by flows converging on leading cities, as has happened so spectacularly in China. Core–periphery structures also exist at an international scale. In the EU a core region encompasses southern England, northern France, Belgium and much of western Germany. Because the EU is founded on a belief in the free movement of labour, huge volumes of economic migrants are leaving the periphery in southern and eastern Europe and heading for the 'cores of the core' – leading cities such as London, Paris, Brussels and Berlin.

Figure 8.9 shows global migration patterns.

Now test yourself

TESTED

15 What is the new global pattern of labour demand?

Answer on p. 222

Economic migrant: A person who travels from one country to another in order to find work and improve their standard of living.

Core–periphery model: This relates to the uneven distribution of population and wealth between two regions and the resulting flows of migrants, trade and investment from a lagging peripheral region to a prospering core region.

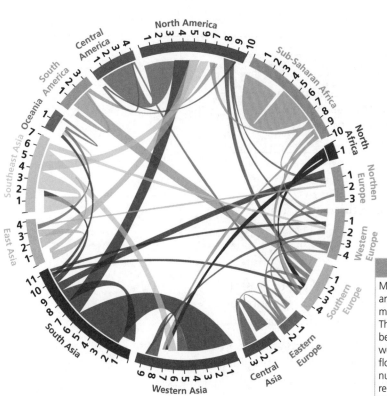

Figure 8.9 Global migration patterns

Immigration policies

As an indicator of the rising tide of international migration, it is estimated that around 4 per cent of the global population lives outside its country of birth. But this proportion varies between countries because of different policies relating to international migration – compare the strict immigration policies of Australia and Japan with the more liberal policies of Singapore and Sweden.

Non-economic migration

Employment is not the only factor in international migration, with unemployment providing the 'push' and job opportunities the 'pull'. There are two other noticeable groups of international migrant, both responses to push forces and therefore forced:

- political **refugees** and **asylum seekers**: the victims of persecution and conflict, such as prevail in the Middle East today
- environmental refugees: where life support systems are collapsing due to climate change and overpopulation, as for example in the Sahel.

Revision activity

From Figure 8.9 make notes about the international migration characteristics of i) North America and ii) South Asia.

Refugee: Defined by the UN as someone whose reasons for migrating are genuinely to do with fear of persecution or death.

Asylum seeker: A person who seeks to gain entry to another country by claiming to be a victim of persecution, hardship or some other compelling circumstance.

Now test yourself

TESTED

16 How does a political refugee differ from an environmental refugee?

Answer on p. 222

The complex causes of migration

> **Key concept**
>
> Migration is the outcome of **push–pull forces** – 'pushing' in the place of origin and 'pulling' in the destination. In the case of voluntary migration, it is likely that the pull factors are stronger, while in the case of forced migration the balance is reversed.

Main causes

The causes of international migration fall into three main groups:

- economic
- political (persecution and conflict)
- environmental.

The risks that huge numbers of migrants have been taking to cross the Aegean and Mediterranean into Europe testifies to the desperation of both economic migrants and refugees to escape from their homelands.

Most international migrants move for work-related reasons and are often driven by poverty and deprivation. For these people, the dividing line between voluntary and forced migration may not always be clear.

Also involved in these work-related migrations are the families who often follow an economic migrant once they have established themselves in the destination country.

Economic theory

The core–periphery model states that with development there is a **backwash** (flows) of migrants to the core from the periphery – a response to the greater opportunities available in the core (Figure 8.10).

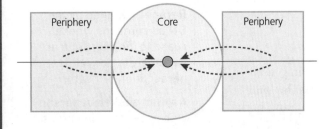

A strong economic core develops fuelled by the in-migration of people (workers and investors) from the peripheral regions of a state.

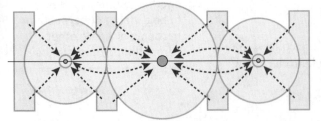

In economic theory, additional core regions form as part of the development process over time. The growth of these cores is fuelled by flows of raw materials and workers from neighbouring areas.

Time

Figure 8.10 Backwash processes in the core–periphery model

Periphery to core migration within a country is not without its problems. For example, there is likely to be some political concern for the plight of the periphery. But when cores reach such a scale and magnetism that they draw in large numbers of foreign migrants, the situation can become much more serious. Such flows can seriously challenge **national identity** and **sovereignty** (see below).

> **Exam tip**
>
> The core–periphery model is important in understanding the geographical distribution of development and other aspects of globalisation. Be sure you have a sound grasp of the basics.

Now test yourself

17 What are the downsides to the growth of large core regions?

Answer on p. 222

Unrestricted flows

There is a strong divergence of opinion about the unrestricted movement of migrants. There are those who argue that economic efficiency is maximised when goods, capital and labour can move freely across international borders. But there are those who focus on the problems of overheated and overgrown cores and the plight of debilitated peripheries.

The free movement of labour is one of the founding principles of the EU and made even easier by the Schengen Agreement. However, the UK is not the only member state concerned about the great influx of migrants involving both economic migrants and refugees. Immigration can reach such a pitch that countries feel they can no longer cope in providing housing, education and healthcare.

> **Revision activity**
>
> Check what the Schengen Agreement is about.

The consequences of international migration

REVISED

Migration has become one of the key issues of the twenty-first century. Because of globalisation and a shrinking world, migration is now a major management challenge for most governments, particularly in those developed countries which are attracting large flows of immigrants.

Now test yourself

TESTED

18 What is meant by a shrinking world and what is its significance?

Answer on p. 222

Immigration changes the **culture** and **ethnicity** of a population. The history of the UK since around 1950 provides a prime example. Despite much **assimilation**, the UK today has become a truly multicultural and multi-ethnic society. Less than 75 years ago it was overwhelmingly a white society. Diversity as such was generated by influxes of Jewish and Irish immigrants.

> **Ethnicity:** Refers to social groups distinguished by their religion, language, customs and heritage.
>
> **Assimilation:** The process by which persons of diverse cultural and ethnic backgrounds interact and intermix in the life of the larger community or nation.

> **Exam tip**
>
> Some people assume that culture is part of ethnicity; others believe otherwise. The specification seems to separate the two. So be warned!

Political tensions

The paradox of globalisation is that while it promotes a global way of thinking, it simultaneously ignites fears that immigrants pose a destabilising threat. The perception prevails that they are taking over housing, jobs and businesses. Almost inevitably, tensions are generated between the natives and the newcomers. Most times, the tensions take the form of ill-informed suspicions rather than outright conflict or hostility.

A good example is provided by the USA and labour flows from neighbouring Mexico. Heightening the tension here is the fact that huge numbers of Mexicans are entering the USA illegally. This raises the suspicion that they are not paying taxes on their earnings.

An uneven playing field

The world's poorest people are sometimes the least likely to become economic migrants. This is because those countries with tight border controls are in a position to select whom they admit. They can choose on the basis of skills that the country needs, as with Australia. Money and influence can also be used to acquire necessary visas. This sort of discrimination is often used as an argument in favour of open national borders.

In the case even of refugees, there is also an element of discrimination. The first to escape tend to be the healthy, young adults and the better informed. The sick, elderly and those with a poor awareness of the threat hanging over them tend to lag behind and often until it is too late.

> **Typical mistake**
>
> It is wrong to assume that all migrants are poor people.

Now test yourself

TESTED

19 What are the main arguments for and against open borders?

Answer on p. 222

The evolution of nation states in a globalised world

> **Key concepts**
>
> **National identity** is a sense of a **nation** being a cohesive whole as a result of its distinctive traditions, culture and language.
>
> **Sovereignty** is the ability of a place and its people to self-govern without any outside interference.

A state is a territory over which no other holds power or **sovereignty**. The term **nation** refers to a group of people who live in a particular territory but who may lack sovereignty. So the Scottish and Welsh nations are part of the UK, which is a sovereign state.

States and the processes that shape them

REVISED

Nation states vary in their unity

Some states pride themselves on their homogenous culture which they have maintained over long periods of time. Examples would include Iceland (because of its isolated location) and North Korea (because of its self-imposed political isolation). Other states boast about their cultural and ethnic diversity, for example Australia and the USA.

In virtually all democracies, national unity is tested by political parties, particularly at the time of general elections.

The nature of borders

Prior to today's shrinking world, mountain ranges and sometimes rivers formed natural barriers to population movements. As a consequence, they provided obvious and quite effective borders. Inside those borders, the isolated conditions favoured the gradual development of a distinctive culture and **national identity**. Of course, the sea has always been an effective border. This explains why many national states are island states.

In Europe, the political map corresponds broadly with the cultural and linguistic maps. In Africa, however, the situation is very different. Political boundaries, drawn centuries ago by colonial powers, do not correspond well with the distributions of different cultural (tribal) and ethnic groups. This explains much of the conflict that has characterised post-colonial Africa.

Exam tip

It is worth knowing one or two examples of states where the boundaries cut across cultural or ethnic divides.

Contested borders

It is not only in Africa where those contested borders dissecting tribal homelands have generated conflict. Other examples include:
- the wars that have waged in Europe between France and Germany over possession of the mineral-rich regions of Alsace and Lorraine
- the boundaries of Russia, which have changed a great deal, most recently in 2014 when Russia annexed part of the Ukraine
- the unhappy fit in the Middle East between state borders and the distribution of strongly opposed ethnic groups.

Another issue here is that of non-recognition. The UN recognises 196 states, but two are not universally recognised as sovereign states:
- Kosovo: it broke away from Serbia in 2008, but it is not recognised by Russia and Serbia
- Taiwan: before the Second World War Taiwan was part of China. It broke away when the communists overran mainland China. Needless to say, Taiwan is not recognised by China as a sovereign state.

Nationalism

REVISED

Many people today view themselves as being global in their outlook. In contrast, others continue to view **nationalism** as a better option.

Empires

In the nineteenth century nationalism was important in:
- the maintenance of empires that European powers had built up earlier as they colonised Africa, Asia and Latin America. The British Empire was the greatest in extent.
- fuelling conflict in Europe, as between France, Germany and Britain.

Nationalism: The belief held by people belonging to a particular nation that their own interests are much more important than those of people belonging to other nations.

New nation states

The era of empire started to come to a close in the late nineteenth century and between 1945 and 1970 the remaining colonised nations gained their freedom. A raft of new nation states emerged.

The rapidity of decolonisation often left a 'power vacuum' and the transition to independence did not bring development. In many countries, power was seized by the army, for example in the Congo, Nigeria and Indonesia. In others, old tribal and ethnic tensions resurfaced, as in Rwanda and Uganda. In others, the handover of power was distinctly messy, as in Kenya, Malaysia and Vietnam.

Colonial heritage

Interesting relics or reminders of the former colonial empires are the flows of migrants between former colonies and the imperial core country. Figure 8.11 illustrates important post-colonial migrations into the UK between 1950 and 1980. Most, but not all, the flows of migrants into the UK were primarily 'pulled' by the job opportunities created by the UK's booming economy and its shortage of labour. Noteworthy exceptions were the refugee movements 'pushed' by persecution and discrimination.

These immigrations were the start of a process that has profoundly affected the ethnicity and cultural diversity of the UK's population today.

> **Typical mistake**
>
> Not all immigrants entering the UK are economic migrants.

> **Revision activity**
>
> Make notes on the origins of the immigration flows shown in Figure 8.11.

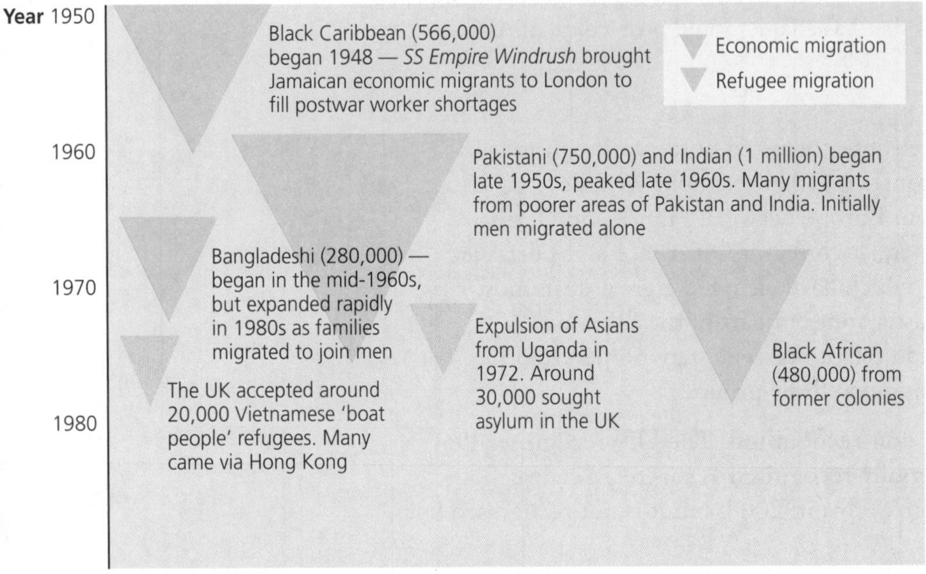

Year 1950

Black Caribbean (566,000) began 1948 — *SS Empire Windrush* brought Jamaican economic migrants to London to fill postwar worker shortages

▼ Economic migration
▼ Refugee migration

1960

Pakistani (750,000) and Indian (1 million) began late 1950s, peaked late 1960s. Many migrants from poorer areas of Pakistan and India. Initially men migrated alone

Bangladeshi (280,000) — began in the mid-1960s, but expanded rapidly in 1980s as families migrated to join men

1970

Expulsion of Asians from Uganda in 1972. Around 30,000 sought asylum in the UK

Black African (480,000) from former colonies

1980

The UK accepted around 20,000 Vietnamese 'boat people' refugees. Many came via Hong Kong

Figure 8.11 UK post-colonial immigration, 1950–1980

Now test yourself

TESTED ☐

20 Why were there labour shortages in the UK and what types of job did the immigrants fill?

Answer on p. 223

Globalisation and new state forms

Globalisation has led to the deregulation of capital markets. This has created an opportunity for some of the newcomers to the ranks of global states to adopt a new way of making a living.

Tax havens

One of the most popular of these opportunities has been the creation of low-tax regimes, which have provided havens for business corporations and wealthy people. Such states include the Bahamas, Gibraltar, Bermuda, Switzerland, the Cayman Islands, the British Virgin Islands, Hong Kong, the Channel Islands and some US states (for example, Nevada and Wyoming). These havens are popular with wealthy **expatriates** and TNCs that are in the business of **transfer pricing**. The latter can be elusive when it comes to paying tax. It is difficult to identify in which country or countries they should pay the taxes on their profits (Figure 8.12).

> **Expatriate:** Someone who has migrated to live in another state but remains a citizen of the state where they were born.
>
> **Transfer pricing:** A financial flow occurring when one division of a TNC charges a division of the same company based in another country for the supply of a product or service. It can lead to less corporation tax being paid.

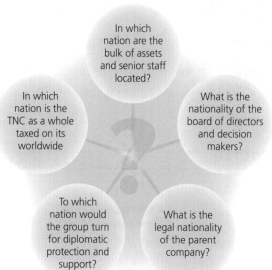

Figure 8.12 Investigating where a TNC is domiciled

Objections

The existence of these tax havens is not popular with many states which desperately need all the corporate taxes they can collect. These taxes are vital sources of revenue. Non-government organisations, particularly those connected with aid, are among those unhappy about the havens. NGOs also see the tax havens as depriving them of much-needed funding.

Alternative models

The present global economic system is far from perfect. It has been frequently criticised by the leaders of several Asian, African and South American nations. Their criticisms centre on the inequalities in the system whereby some states gain disproportionately from global trade at the expense of others.

Unfortunately, the governments that most criticise global capitalism have yet to persuade the rest of the world that there exists a viable alternative form of state building. The question arises: would you really like to live in North Korea, Zimbabwe, Ecuador or Bolivia?

Revision activity

Why have the tax affairs of companies such as Amazon, Apple, Google and Amazon been criticised?

Exam tip

It is important to realise that globalisation is not the silver bullet solution to creating a more even global spread of development.

Now test yourself   TESTED

21   In which of the four countries – North Korea, Zimbabwe, Ecuador or Bolivia – would you prefer to live? Give your reasons.

Answer on p. 223

The impacts of global organisations

Since the end of the Second World War in 1945, there has been an acceleration towards **global governance**. Leading the way was the United Nations (UN), an umbrella organisation for many global agencies, treaties and agreements.

Global organisations in the post-1945 world

The United Nations

The UN was the first post-war IGO to be set up. Since 1945, its remit has grown to include a whole range of different areas of governance (see 'Interventions', below).

The **Security Council** is an important and powerful arm of the UN. Its primary responsibility is the maintenance of international peace and security. Five powerful states (China, France, Russia, the UK and the USA) are permanent members. In recent years, its record of successful interventions has been somewhat tarnished by divisions within the Council (see below).

Interventions

The main interventions made by the UN fall under the following headings:

- direct military intervention – for example, in the Congo
- peacekeeping in states with internal conflicts – for example, Ivory Coast
- economic sanctions against countries stepping out of line – for example, trade embargo on Iran
- defending human rights – for example, setting up war crime trials
- protecting refugees – for example, in Syria
- promoting development, especially in agriculture, education and healthcare – for example, over much of Africa.

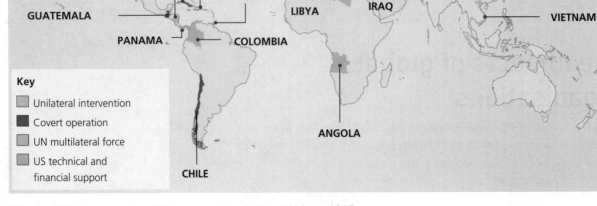

Figure 8.13 Some major US interventions abroad since 1945

Unilateral actions

Despite recognising the UN, some member states have conducted their own military campaigns and contrary to the resolutions of the Security Council. Examples include:

- the UK's war in 1982 with Argentina over the sovereignty of the Falkland Islands
- the US invasion of Iraq in 2003, with the support of the UK
- Russia's annexation of Crimea in 2014.

Since 1945, there have been many examples of unilateral military interventions in so-called **failed states**, such as Yemen, Somalia and Syria, or on the pretext of waging a war on international terrorism, for example in Afghanistan.

> **Typical mistake**
>
> It is wrong to think that the power of the UN is paramount.

> **Failed state:** A country whose government has lost political control and is unable to fulfil the basic responsibilities of a sovereign state, with severe adverse impacts on some or all of its population.

Now test yourself

 TESTED

22 Why has it been necessary for the UN and its agencies to make interventions?

Answer on p. 223

The control of trade and financial flows

REVISED

Three powerful players

Three IGOs – the World Bank, the IMF and the WTO – have been particularly significant in the management of the global economy (Table 8.5). Established in the 1940s, they have since become important in maintaining the global dominance of Western capitalism.

Table 8.5 Three global players in economic development

Organisation	Founding date	Headquarters	Role in world trade
World Bank	1944	Washington DC, USA	To give advice, loans and grants for the reduction of poverty and the promotion of economic development. The World Bank's main role is to offer long-term assistance rather than crisis support.
IMF	1944	Washington DC, USA	To monitor the economic and financial development of countries and to lend money when they are facing financial difficulty. Help is provided to countries across the development spectrum. Between 2010 and 2015, almost US$40 billion was lent to Greece, for instance.
WTO (previously GATT)	1995 (previously 1947)	Geneva, Switzerland	To formulate trade policy and agreements, and to settle disputes. Overall, the WTO aims to promote free trade on a global scale. Unfortunately, a round of negotiations that began in 2001 stalled for 14 years. Trying to get 162 member states to agree anything can be challenging. Difficult problems for the WTO to deal with include: • wealthy countries failing to agree on how far trade in agriculture should be liberalised • the fast growth of emerging economies including China (which makes it harder to agree on fair policies for so-called developing countries).

Controversial borrowing rules

Lending by the IMF and the WTO has undoubtedly helped many lagging states to develop economically. However, since 1970 tougher conditions have been attached to large-scale lending. For states experiencing severe financial difficulties, there are two forms of help available:

- structural adjustment programmes (SAPs): providing loans but with strict conditions attached
- heavily indebted poor countries (HIPC) policies: aimed at ensuring that no poor country faces a debt burden it cannot manage.

Critics of these two concessions say that they tend to increase rather than decrease poverty. The concessions also undermine the economic sovereignty of borrowing states. They are regarded by some as a neo-colonial strategy used by developed countries to maintain influence over how peripheral countries develop.

Regional groupings

As a reaction to the failure of the WTO to deliver free trade globally, states have organised themselves into regional groupings or **trade blocs** (Figure 8.14).

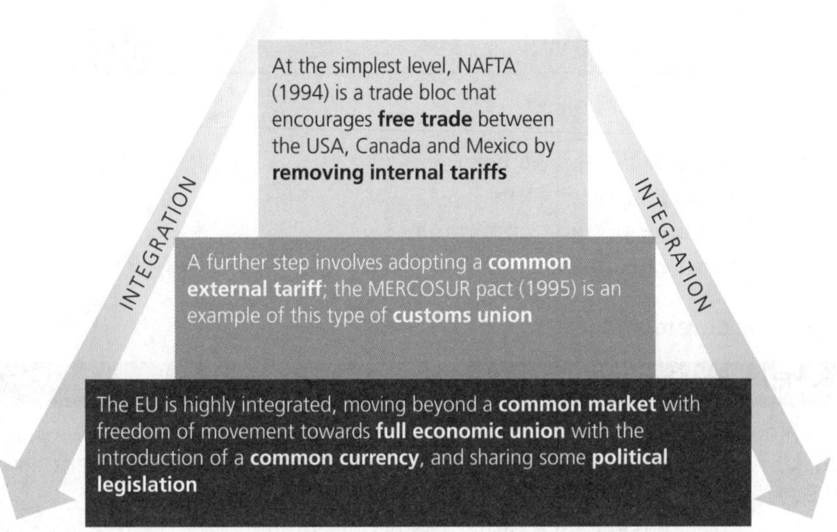

At the simplest level, NAFTA (1994) is a trade bloc that encourages **free trade** between the USA, Canada and Mexico by **removing internal tariffs**

A further step involves adopting a **common external tariff**; the MERCOSUR pact (1995) is an example of this type of **customs union**

The EU is highly integrated, moving beyond a **common market** with freedom of movement towards **full economic union** with the introduction of a **common currency**, and sharing some **political legislation**

INTEGRATION

INTEGRATION

Figure 8.14 Different types of trade bloc and their degree of integration

> **Exam tip**
>
> It is recommended that you have some information about NAFTA or ASEAN, MERCOSUR, and the EU.

Global environmental governance

REVISED

The UN has attempted to manage some of the world's pressing environmental issues, but with varying degrees of success.

Global environmental issues

The atmosphere and the biosphere are important resources shared by all members of the global community. Table 8.6 sets out important agreements for each of the two spheres.

Table 8.6 Global agreements and actions on the atmosphere and the biosphere

Agreements and actions	UN aims	Evaluation
Montreal Protocol on Substances that Deplete the Ozone Layer	In the late 1960s the UN Environment Programme first called for an international response to the issue of ozone depletion caused by worldwide use of a group of chemicals called chlorofluorocarbons (CFCs) in fridges and aerosol sprays. The 1977 'World Plan of Action on the Ozone Layer' gave the UNEP responsibility for promoting and coordinating international research and data-gathering activities.	The Montreal Protocol was signed in 1987. It was a remarkable agreement on account of the number of individual governments that were prepared to back an important global goal ahead of narrower economic self-interest. Within a decade, irrefutable proof that CFCs were to blame led to all UN states ratifying the treaty. CFC use was phased out rapidly as a result of this exceptional international cooperation.
Climate change agreement	Climate change was first raised as an urgent issue in 1992 at the UN Earth Summit conference. Many of the meetings that followed were plagued by uncertainty over the evidence and also wrangling over which nations should be held responsible for the majority of the anthropogenic carbon stock that has been added to Earth's atmosphere.	On the whole, international cooperation on climate change has taken place very slowly, which many people deem to be a failure of international governance. Although a new international agreement on action was reached in Paris in 2015, critics say that the pledges that were made to reduce carbon emissions do not go far enough. These pledges cannot be enforced either.
Convention on International Trade in Endangered Species of Wild Fauna and Flora (CITES)	CITES entered into force in 1975. It banned trade in threatened species and their products. Now adopted by 181 countries, it has effectively saved some species but not others. Success stories include recovery of the Hawaiian nēnē bird and the Arabian oryx.	Rising wealth in China, Indonesia and South Korea has actually increased illegal trade in some prohibited substances such as ivory and rhino horn. The problem can be summed up as 'new money, old values'. Without a cultural shift away from the use of these products, CITES will not be able to protect some species.
Millennium Ecosystem Assessment	This international collaboration helped popularise the 'ecosystem services' approach to biodiversity management. A financial value is calculated for threatened biomes and species, thereby strengthening the rationale for their preservation.	'Ecosystem services' is a philosophy that fits well with the capitalist values of the global economic system and is a pragmatic approach for the UN to have adopted and helped promote globally.

Now test yourself

TESTED ☐

23 Which of the four aims in Table 8.6 do you think is the most challenging? Give your reasons.

Answer on p. 223

Revision activity

Make brief revision notes about all four agreements in Table 8.6.

Global waters

IGOs have been developing laws for managing the world's oceans and international rivers:

- UN Convention on the Law of the Sea (launched 1982): defined a nation's territorial limit as extending to 12 miles and its exclusive economic zone to 200 miles from the coastline. It recognises that oceans are **global commons** and sets out laws to protect the marine environment from pollution.

Global commons: Global resources so large in scale that they lie outside of the political reach of any one state. International law identifies four global commons: the oceans, the atmosphere, Antarctica and outer space.

● Helsinki Water Convention (launched 1992): aims to protect and ensure the quantity, quality and sustainable use of transboundary water resources by facilitating cooperation between the states involved.

Now test yourself

Now test yourself

TESTED

24 What are 'international rivers'? Give three examples.

Answer on p. 224

Managing Antarctica

Antarctica, the world's fifth largest continent, is now universally recognised as a 'continent of peace and science'. Recognition of its intrinsic value and special status started in 1959 with the Antarctic Treaty. Nobody owns Antarctica, but around 20 nations have permanent scientific bases there.

For the immediate future, Antarctica is safe from exploitation, except possibly by tourism. Around 40,000 tourists visit Antarctica in cruise ships each year. However, as non-renewable resources run out elsewhere in the world, it might be foreseen that there will be increasing calls to exploit the continent's coal, oil, copper, silver, gold and titanium.

Now test yourself

TESTED

25 How do tourist cruises threaten Antarctica?

Answer on p. 223

Threats of national sovereignty in a more globalised world

There is one theory (hyperglobalisation) which states that globalisation will, over time, reduce the power of individual countries. Global flows of commodities, ideas and people in a shrinking and borderless world will gradually lead to the creation of a global village where group identities will give way to **global citizenship**. In contrast, others argue that there is a growing resistance to globalisation and membership of regional groupings. In many states and nations attempts are being made to reassert national identity.

> **Global citizenship:** A way of living in which a person identifies strongly with global-scale issues, values and culture rather than with a particular nation, state or place.

The elusive concept of national identity

REVISED

> **Key concept**
>
> **National identity** is an elusive concept in terms of trying to identify and measure it. Part of the problem lies in the fact that it is the product of many interacting things, some of which are intangible (for example, the fact that national identity means different things to different people adds to the difficulty). Perception is paramount. What we can be certain about is that national identity, and the wish to preserve it, lies at the heart of nationalism.

Factors reinforcing nationalism

There is evidence in some countries of a growing concern about the amount of sovereignty being surrendered as a result of participation in globalisation. The feeling is that governance is now in the hands of the UN and IGOs rather than national government. Some of the support for the UK's decision to exit from the EU came from the conviction that the UK was being increasingly ruled from Brussels and not London. Loss of autonomy equals loss of sovereignty.

There are other non-political factors encouraging nationalism, for example:
- education that informs about national heritage
- prestige in the world of sport – for example, the Olympic medal tables
- the pride that individuals feel in belonging to a particular nation
- pride that comes when citizens are renowned internationally, from pop stars to Nobel prize winners
- pride that so many overseas tourists wish to visit.

Identity, loyalty and national character

Clearly, these three aspects are closely interrelated and important components of national identity. Aspects of national identity are sometimes tied to distinctive legal systems and methods of government. For example:
- Americans' belief in the First Amendment
- the legacy of the French Revolution and the importance of *liberté* in modern France
- the UK's Magna Carta and its foundation of British laws and liberties.

Multinational countries

One of the features of most countries today is their increasingly multinational nature. This is largely the outcome of recent international migration flows. The UK is a prime example. The question here is this: does this increasing ethnic and cultural diversity strengthen or weaken national identity?

Table 8.7 How English national identity has changed over the past 100 years

	Early twentieth century	Twenty-first century
Religious beliefs	Generally widespread, with high levels of Anglican or Catholic church attendance	Largely secular and non-religious, although some minority faiths are prospering
Food	Locally sourced seasonal food; native herbs preferred to foreign spices	Global, varied tastes in food; strong spices are widely used in cooking
Identity	People had a strong sense of local belonging (either to a town or county); regional dialects were stronger than today; most were also extremely patriotic and would fight for their country	Many would be less willing to fight for their country, although they are often strong supporters of national football teams; younger people may see themselves as 'global citizens'
Roots of vocabulary	Celtic, Saxon, Scandinavian (Norse), Roman, Greek, French	Additional Indian, Jamaican and American influences (due to migration and TV)

Now test yourself

TESTED

26 Can you suggest one explanation for the changes in English national identity shown in Table 8.7?

Answer on p. 223

Challenges to national identity

REVISED

National identity is being challenged in a number of ways. Three examples follow.

Foreign-owned companies

Many UK companies are foreign-owned – for example, EDF, Jaguar and Land Rover. For this reason, the term 'Made in Britain' does not quite mean the same as it did say 50 years ago. So a possible ingredient of national identity has been lost.

Westernisation

Westernisation, which is spreading as a sort of global culture, is dominated by US cultural values. Some would say that Westernisation equals Americanisation. No matter what it is called, it is being promoted by large retailing and media corporations (Table 8.8). There can be little doubt that the spread of Westernisation, with its particular set of capitalist values, is eroding some of the differences between nation identities.

Table 8.8 Three large entertainment companies that contribute to 'Westernisation'

Walt Disney Company	This is the world's largest entertainment TNC with annual earnings exceeding US$50 billion. Across Asia, young children encounter Mickey Mouse and similar brands on Disney Channel Asia. In Disney movies and magazines, they are also exposed to Western traditions including Christmas and Halloween. Disney is not deliberately trying to change cultures in other countries, but this is often the result.
MTV	The company has adopted a '360-degree strategy' of global marketing with an extensive network of TV stations now delivering Anglo-American music to every continent, including Africa. Pop music spreads many Western cultural traits, including American fashion and the English language. By promoting strong female artists, MTV may also be promoting greater gender equality.
Apple	The apps that are bundled up with an iPhone promote Western culture in all kinds of subtle ways; St Valentine's Day may be mentioned in a calendar, for instance.

Ownership of property

The non-national ownership of property is perceived by many as also threatening national identity. Examples include:
- Russian and other foreign investment in the London property market as a safe haven for the unstable currencies – note the particular concentrations in Figure 8.15
- UK retirees in Spain purchasing residential properties
- US and Indian ownership of TNCs.

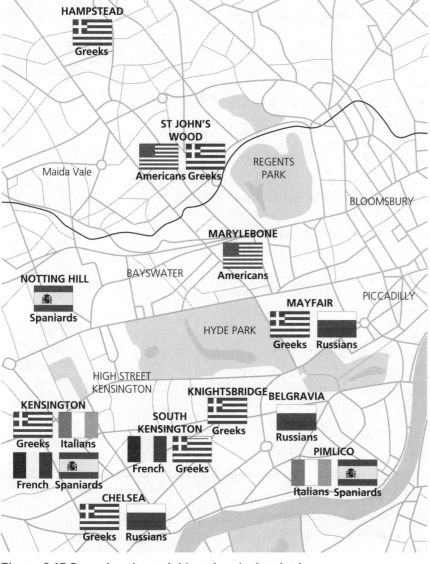

Figure 8.15 Some London neighbourhoods developing a strong non-British identity

Consequences of disunity within nations

REVISED

The distinction between nation and state has already been drawn (see page 192). The term nationalism is used in two different contexts:

- the promotion of sovereign states, such as the UK
- the promotion of small nations which lack sovereignty but are part of a sovereign state, such as Scotland and Wales in the UK.

Nationalist movements

In recent years there has been an upsurge in nationalism of the latter type. Examples include the Scottish Nationalist Party's bid for independence from the UK, the Catalonian wish to gain full independence from Spain, and the break-up of former Yugoslavia in the early 1990s, which gave many small nations (such as Slovenia, Croatia and Montenegro) the chance to assert their sovereignty.

An interesting question is this: is this upsurge in nationalism really a reaction to globalisation? Certainly there are those who see globalisation as eroding nationalism and promoting global citizenship.

> **Exam tip**
>
> With this sort of contentious question, you will impress the examiner if you show that you are capable of seeing both sides of it.

Political tensions within the BRICs

Many emerging economies, including members of the BRIC group, are experiencing significant internal political tensions:

- Brazil: polarisation of rich and poor; corruption
- Russia: ethnic discrimination; power of political elite
- India: polarisation of rich and poor
- China: tensions between rural and urban communities; human rights issues.

Failed states

National unity and national identity are most challenged in the so-called **failed states**. Examples include Rwanda, Somalia and Sudan in Africa, and Iraq, Yemen and Afghanistan in Asia. Here governments have lost control and political and economic power lies with wealthy elites, foreign investors, terrorist organisations or a dominant and powerful ethnic or cultural group.

The issue raised by the failed states is a thorny one. Should they be left alone to heal their own disunities? Or should there be some form of intervention? If the latter, who should intervene, the UN or some superpower with its own agenda?

Now test yourself

TESTED

27 Give some possible causes of disunity within nations.

Answer on p. 223

Skills reminder

You should be familiar with the skills and techniques used in the following investigations of migration, identity and sovereignty:

- using flow lines on global maps to show migration movements
- interpreting migrant oral accounts
- interpreting a range of opinions about the contributions made by migrants
- using divided bargraphs to compare the ethnic diversity of countries
- comparing maps of languages and colonial histories
- using the Gini coefficient and other techniques to measure inequalities of income and wealth
- evaluating sources to determine the impact of IGOs managing global environmental issues
- using proportional circles to show the output and level of foreign ownership of different economic sectors
- critical analysis of source materials used in the assessment of the costs and benefits of foreign ownership
- critical analysis of source materials used in the assessment of attempts to promote national identity.

Exam practice

A-level

1 (a) Study Figure 1.

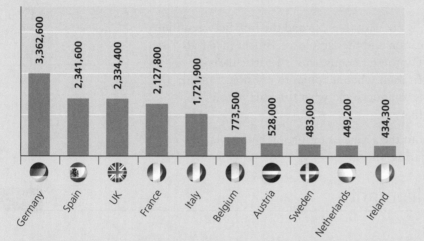

Figure 1 EU states with the largest numbers of people born in another EU state, 2012

(i) Explain how this analysis of international migration within the EU might have been improved. (3)

(ii) Name a technique widely used in the analysis of migration. (1)

(b) Explain what is meant by **national identity**. (6)

(c) Explain why the pattern of international migration is changing. (8)

(d) Evaluate the reasons why nationalism remains a powerful force today. (20)

Answers and quick quiz 8B online

ONLINE

Summary

You should now have an understanding of:
- globalisation and the changing pattern of labour demand and migration
- the causes of migration
- the varied consequences of international migration
- the challenges posed by migration
- nation states and their diverse histories
- nationalism and its role in the modern world
- the deregulation of capital markets and the emergence of new state forms

- the importance of IGOs in today's world
- the IMF, the World Bank and the WTO and the maintenance of Western capitalism
- IGOs and the management of environmental problems
- the elusive nature of national identity
- the challenges to national unity
- the consequences of disunity within nations
- failed states and their internal tensions.

9 Synoptic themes

This part of the specification is not easy to grasp and the link between the content set out in the specification and the sample assessment materials is not immediately obvious. For these reasons, this is a particularly challenging part of the qualification when it comes to revision. Indeed, perhaps the best you can do is to understand what this part of the qualification is seeking to achieve.

It is important to be clear about the actual examination before getting down to the business of preparing for it.

Examination requirements

AS examination

In Paper 1, there will be a compulsory synoptic theme question in the section relating to your chosen options: *either* **2A Glaciated landscapes and change** *or* **2B Coastal landscapes and change**.

In Paper 2, again there will be a compulsory synoptic theme question relating to your chosen option: *either* **4A Regenerating places** *or* **4B Diverse places**.

All questions make reference to materials in a separate resource booklet.

In short, you will have to attempt two synoptic theme questions. So this is a significant element in the overall assessment (approximately 18 per cent of the qualification).

A-level examination

Paper 3 is devoted entirely to the synoptic themes. All six questions are linked to maps, diagrams, tables, etc. provided in a resource booklet.

This paper accounts for 20 per cent of the qualification.

Synoptic themes (PAF)

The term is misleading in that what the specification sets out are not linking topics and concepts but rather three different perspectives from which to look at some of the important issues of today. These have been flagged throughout this book using synoptic theme boxes.

Players (P): they are individuals, groups, organisations, governments, businesses and other stakeholders involved in geographical issues and decisions in a number of different ways:
- as **causes** of or contributors to the issue, for example the role of refugees in the so-called migrant crisis
- as the **victims**, both directly and indirectly, of the issue, for example the role of local residents affected by an airport expansion
- as **managers** making decisions about how best to deal with the issue, for example governments dealing with energy security.

It is important to note that some players have greater influence than others.

Exam practice answers and quick quizzes at **www.hoddereducation.co.uk/myrevisionnotes**

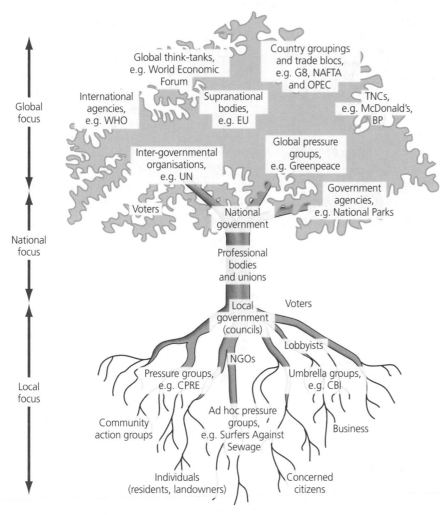

Figure 9.1 Players in issues decision making

Figure 9.1 shows the range of potential decision makers on three spatial scales: global, national and local.

Attitudes and actions (A): all the players involved in an issue will have their own attitudes or views on it. The attitudes of different players may well clash. In some instances, they may have taken actions which they thought were appropriate or necessary. Influences on attitudes include identity, political and religious views, and profits, as well as importance attached to social justice, equality and the natural environment.

Futures and uncertainties (F): this concerns rather more those players who make decisions that are likely to affect people in the future. There are three different approaches to management of the future:
- opting for business as usual
- seeking sustainable strategies
- devising radical alternatives (mitigation and adaptation).

The links between the three standpoints are shown in Figure 9.2.

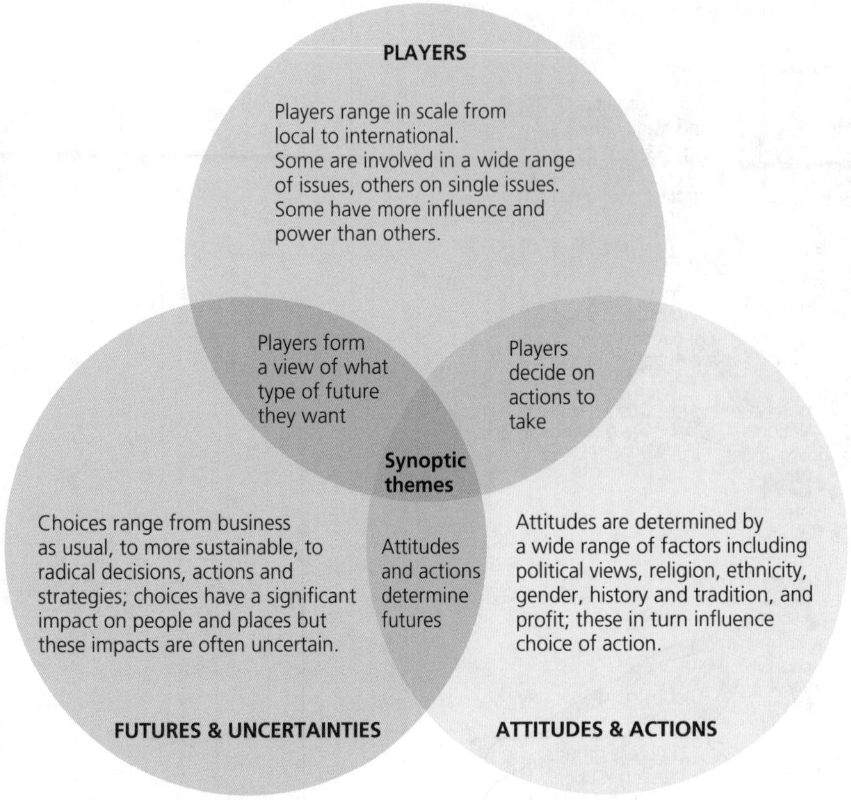

PLAYERS

Players range in scale from local to international.
Some are involved in a wide range of issues, others on single issues.
Some have more influence and power than others.

Players form a view of what type of future they want

Players decide on actions to take

Synoptic themes

Choices range from business as usual, to more sustainable, to radical decisions, actions and strategies; choices have a significant impact on people and places but these impacts are often uncertain.

Attitudes and actions determine futures

Attitudes are determined by a wide range of factors including political views, religion, ethnicity, gender, history and tradition, and profit; these in turn influence choice of action.

FUTURES & UNCERTAINTIES

ATTITUDES & ACTIONS

Figure 9.2 The links between the three synoptic standpoints

Whatever strategy is adopted, there will be positive and negative impacts, winners and losers. There will also be consequences that relate to risk, resilience and thresholds. But can we really be sure that what we decide now will have the desired outcome tomorrow? So in anticipating any futures, there are uncertainties: scientific, demographic, economic and political. These uncertainties relate to:

- the accuracy or otherwise of forecasts and predictions
- unforeseen events and changes
- possible outcomes.

Unfortunately, experience teaches us that forecasts and predictions can go horribly wrong.

Which of these three synoptic perspectives to adopt on an issue is clearly signalled in the preceding topics in the synoptic theme boxes. Table 9.1 gives examples from each of the topics.

Table 9.1 Some examples of the so-called synoptic themes

Topic	Theme	Sample players (other than national and local government)	Attitudes and actions	Futures and uncertainties
1	The hazard management cycle	IGO; NGOs; local residents	Contrasting attitude to hazard risks	Forecasting disaster impacts
2A	Exploiting glaciated landscapes	Tourism; TNCs; conservationists	Actions and their impact on natural systems	Impacts of climate change
2B	Exploiting coastal landscapes	Developers; tourism; conservationists	Actions and their impact on natural systems	Impacts of climate change

Topic	Theme	Sample players (other than national and local government)	Attitudes and actions	Futures and uncertainties
3	Promoting globalisation	IGOs; NGOs; TNCs	Contrasting attitudes to globalisation	Patterns of resource consumption
4A	Regeneration	Investors, businesses, local groups	Contrasting attitudes to regeneration	Success of regeneration plans
4B	Cultural diversity	Local groups; political parties	Contrasting attitudes to places	Change and legacies
5	Shared waters	IGOs; major water users; scientists	Contrasting attitudes to water supply	Rising global water demand
6	Securing energy supplies and pathways	TNCs; IGOs; green parties	Energy consumers' attitudes to the environment	Rising global energy demand
7	Global policing	IGOs; NGOs; green parties	Contrasting attitudes to the reduction of carbon emissions	Future power structures

The difficulty is that in both examinations (AS and A-level) the questions do not use the terms of the three perspectives underlying the synoptic themes, namely 'players', 'attitudes and actions' and 'futures and uncertainties' – they are simply implied.

Synoptic thinking and core concepts

REVISED

Synoptic thinking means thinking like a geographer. In other words:
- seeing the links between the topics you have studied
- making connections between different places and people
- recognising that there are some concepts or topics that link across most of what you have studied.

Core concepts

The specification sets out a number of core concepts which are, in effect, bridging or truly synoptic topics. These concepts require you to bring together your knowledge and understanding of more than one topic. It will be one or more of these concepts, rather than specific references to players, attitudes or futures, that provides the overarching structure for the synoptic theme questions. For this reason, it is important that you understand these core concepts and the topics they might touch.

The core concepts are now explained and illustrated. You should note that the core concepts do not necessarily have links with all the topics. Hence the blanks in some of the following tables.

Causality

Causality is about the relationship between something that happens or exists and the factors or circumstances that cause it (Table 9.2). Two obvious examples are provided by climate change and globalisation. The causes of climate change are largely physical, with a possible human input (e.g. carbon emissions), while the causes of globalisation are largely human, with a possible physical input (e.g. the availability of resources).

Table 9.2 Some examples of 'causality' topics

Topic	Example of causality topic
1	Tectonic hazards
2A	The fragility of glaciated landscapes
2B	Coastal flooding
3	International migration
4A	The need for regeneration
4B	Changing population structures
5	Water insecurity
6	Energy insecurity
7	The rise of superpowers

Systems and feedback

Systems have two different contexts in this specification:
- as a set of interrelated physical parts with inputs and outputs
- as a three-step decision-making mechanism (Figure 9.3).

Feedback occurs when outputs of a system are returned to the system as inputs. The resulting loop creates a cause-and-effect chain, sometimes referred to as a feedback loop.

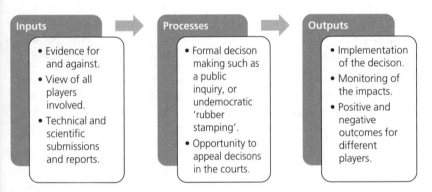

Inputs
- Evidence for and against.
- View of all players involved.
- Technical and scientific submissions and reports.

Processes
- Formal decison making such as a public inquiry, or undemocratic 'rubber stamping'.
- Opportunity to appeal decisons in the courts.

Outputs
- Implementation of the decison.
- Monitoring of the impacts.
- Positive and negative outcomes for different players.

Figure 9.3 The decision-making system

Table 9.3 Some examples of 'systems and feedback' topics

Topic	Example of systems and feedback topic
1	
2A	The glacial mass balance
2B	The coastal cell
3	
4A	
4B	
5	The hydrological cycle
6	The carbon cycle
7	

Inequality

Inequality exists at all geographical scales. It relates to differences in opportunity, access to resources, wealth and influence between different groups. In terms of players, decision-making power and influence tend to be in the hands of people with political power, financial resources or cultural influence.

Table 9.4 Some examples of 'inequality' topics

Topic	Example of inequality topic
1	
2A	
2B	
3	Development gaps
4A	Multiple deprivation
4B	Rural and urban poverty
5	Access to water
6	Access to energy
7	Superpowers and the minnows

Identity

Identity refers to the beliefs, perceptions and characteristics that make one group different from another. This is strongly related to place. Differences in identity mean that we should expect contrasting groups of people not to share the same fears, desires, ambitions or concerns. Identity and a strong attachment to places shape the attitudes of traditional and tribal societies.

Table 9.5 Some examples of 'identity' topics

Topic	Example of identity topic
1	
2A	Indigenous people
2B	
3	
4A	Attachment to place
4B	Attachment to place
5	
6	
7	Political groups

Globalisation and interdependence

Globalisation is the increasing integration of the world's economies, societies and cultures. It is creating a more interconnected and interdependent world. For example, the economic success of one country is tied to, and dependent on, economic success in other countries.

Table 9.6 Some examples of 'globalisation and interdependence' topics

Topic	Example of globalisation and interdependence topic
1	
2A	Pressure on cold environment resources
2B	Development of coastal zone
3	
4A	Impact on regeneration
4B	Impact on places
5	Impact on water demand
6	Impact on carbon cycle and energy
7	

Mitigation and adaptation

Mitigation is preventing something from happening, such as preventing a natural event from becoming a hazard. Adaptation is dealing with the impacts of something, such as building flood defences.

Table 9.7 Some examples of 'mitigation and adaptation' topics

Topic	Example of mitigation and adaptation topic
1	Strategies to modify hazard vulnerability
2A	Protecting permafrost
2B	Coping with future coastal threats
3	
4A	
4B	
5	Desertification
6	Rebalancing the carbon cycle
7	

Sustainability

Sustainability has been a geographical buzzword for some time. More recently, the term 'environmental stability' has entered geographical speak. It implies that people need to reduce their overall impact on the Earth to a sustainable level, one that prevents irreversible environmental damage.

Resources are a major focus in concerns over sustainability. It is here that contrasting attitudes are most obvious, with exploitation at one end of the spectrum and conservation at the other (Figure 9.4).

Protection ← MOTIVE → Profit

Conservation	Sustainable management	Exploitation
Limiting resource exploitation to minimum requirements of humans, and protecting as much of the natural environment as possible	Exploiting resources in some places, conserving in others; where resources are exploited, they are managed to minimise losses to ecosystems and environment	Using natural resources to maximise profits and economic growth; resources are seen in economic value terms only

Figure 9.4 Contrasting attitudes to resources

Table 9.8 Some examples of 'sustainability' topics

Topic	Example of sustainability topic
1	
2A	Threats facing glaciated landscapes
2B	Wetland reclamation
3	Impact on resources
4A	
4B	
5	Managing water resources
6	Managing energy resources
7	Tensions over resources

Risks and thresholds

Risk is best thought of as degree of exposure to harm or the chances of something happening. Risk can relate to avoiding hazards, running out of food, water or energy, contracting a disease or sinking into poverty. Thresholds mark the boundaries between safety and risk. They are tipping points (Figure 9.5).

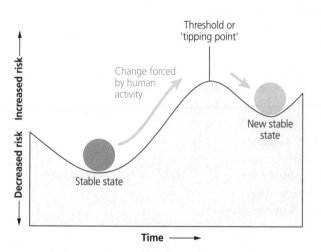

Figure 9.5 The threshold concept

Table 9.9 Some examples of 'risks and thresholds' topics

Topic	Example of risks and thresholds topic
1	The hazard risk equation
2A	Climate warming and unique landscapes
2B	Climate warming and coastal flooding
3	
4A	
4B	
5	Future droughts and floods
6	Continuing use of fossil fuels
7	Geopolitical stability

Resilience

Resilience is the ability to cope with change or stress. Resilient communities can cope with change, both known and unforeseen.

Table 9.10 Some examples of 'resilience' topics

Topic	Example of resilience topic
1	Coping with tectonic hazard events
2A	Coping with resource exploitation
2B	Coping with rising sea level
3	Coping with international migration
4A	Coping with change
4B	Coping with change
5	Coping with water insecurity
6	Coping with energy insecurity
7	Coping with superpowers

Skills

REVISED

The examination questions' synoptic themes often involve interpreting resources, such as maps, diagrams, photographs, etc., and using what they show to support your arguments. So be sure to brush up on your ability to 'read' resources.

Having reminded you of what is involved in those parts of the specification concerned with the synoptic themes, it has to be said that there is little you can do by way of swotting up. The brutal truth is that the examiner can pick any one of the core concepts and pitch their questions at any one or more of your topics. So the possible questions are almost endless. Whatever they turn out to be, it will then be up to you to apply the knowledge and understanding you have gained from those topics. You need to show an ability to address the questions as directly and explicitly as possible. For many of you, this will be the most challenging part of the whole qualification.

Question practice

REVISED

It is recommended that you try your hand at the relevant questions given in the sample assessment materials. Ask your teacher to print off the questions and the relevant parts of the resource booklets. Good luck!

Now test yourself answers

Chapter 1

1 Mediterranean; Caribbean

2 They are 'floating' on the mantle and are driven by currents within it. The outcomes are different types of plate boundary and their associated hazards.

3 Can be argued either way. A high-magnitude event in a sparsely populated location is hardly likely to become a disaster whereas an event of less magnitude in a densely populated area could easily become so.

4 They are waves of energy that travel through the Earth's crust and are a result of earthquakes and volcanic eruptions. The hypocentre is the point of rupture within the Earth where an earthquake starts. The epicentre is the point at the Earth's surface immediately above the hypocentre.

5 Pyroclastic flows because they involve hot, toxic gases and pyroclastic materials. They occur explosively without warning and move downslope at incredible speeds, giving humans little or no chance of escaping.

6 A hazard is an event that threatens both life and property. A disaster occurs when the hazard causes mass destruction and death. The difference lies in the scale of the impacts.

7 Vulnerability is the degree of exposure to the impacts of a hazard. Resilience is the ability of a society or community to cope with and recover from the effects of a hazard.

8 There are many more earthquakes during the course of a year than volcanic eruptions. Earthquakes occur in more parts of the world, including some of the most densely populated. Earthquakes occur without warning, whereas there are often tell-tale symptoms that warn of a forthcoming eruption.

9 They compare the physical characteristics and processes that all hazards share. Comparisons can help decision makers identify and rank the hazards that should be given priority in terms of attention and resources. Hazard profiling is more difficult when comparing across hazards, but can work well when comparing the same type of hazard event.

10 Good governance should ensure the proper use of resources in trying to minimise the impacts of a hazard. So this includes supporting a whole range of actions educating the public about what to do when a hazard strikes, planning emergency procedures, land-use planning, building design and construction, etc.

11 Because earthquake damage to a building, for example, can range from a few cracks to total destruction. The problem for the insurance company is in assessing the scale of risk: will another event occur here, and when? What will be the scale of damage? Insurance is a business, so premiums will always be pitched high to cope with the risk factor (uncertainties) and to safeguard the profit margin.

12 Most criticism focuses on the emergency phase. Trying to put into effect even a rehearsed set of procedures is often very difficult when chaos prevails. It is difficult to coordinate the work of voluntary rescue teams and aid organisations, particularly if they come from overseas and lack first-hand experience of the environment in which the disaster has occurred. The destruction of infrastructure, particularly roads and airports, adds to the challenges of coping in the immediate aftermath of a disaster. Another concern is the misappropriation of funds raised by public appeals across the globe. The Haiti earthquake of 2010 provides a classic example.

Chapter 2

1 Long-term causes are explained by Milankovitch's theory and are to do with the Earth's orbit around the Sun. Short-term climate change is related to sunspot activity, volcanic activity and most recently human activity.

2 Ice cover has been shrinking since about 1850 thanks to a general warming of the global climate. Ice accumulations in mountains are generally thinner anyway because they are feeding warm-based glaciers. The lower the latitude of the high mountains, the greater the reduction simply because of the warmer climate.

3 Aspect determines both the amount of snow falling and the accumulation. In the northern hemisphere, north- and northeast-facing slopes are both more sheltered and shaded. They are therefore more conducive to snow accumulation and the compaction of snow into ice.

4 Permafrost and frost-formed features such as ice-wedge polygons, patterned ground, pingos, block fields and rock glaciers. There are also landscape features resulting from wind (loess) and meltwater rivers (braided channels).

5 The balance between its gains (accumulation) and its losses (ablation). The place where accumulation and ablation balance each other out is known as the equilibrium point.

6 Open, because there are inputs (snow, weathered material) and outputs (meltwater, moraine). Components are solar energy, snow and ice, rock debris, moraine and meltwater.

7 Altitude affects both temperature and precipitation. Higher altitude means lower temperatures and higher precipitation. The former will slow glacier movement; the latter will mean more accumulation and possibly faster movement.

8 Upland glaciated landforms are distinguished from lowland glaciated areas by a suite of landforms produce by erosion – for example, cirques, pyramidal peaks, U-shaped troughs, hanging valleys and truncated spurs.

9 The key features are the cirques. As a result of headwall erosion, when two cirques lie either side of a ridge they cut back towards each other and reduce the ridge to an arête. Where more than two cirques are converging through headwall erosion, the separating upland mass will be reduced to a pyramidal peak.

10 a) A terminal moraine marks the furthest point reached by a valley glacier, while recessional moraines mark still stands in the retreat of a glacier's snout.

b) Lateral moraines are linear ridges deposited along valley sides. Medial moraines develop on the surface of a valley glacier and are formed by the joining of two separate ice streams (i.e. lateral moraines).

11 It is the temperature at which ice melts at a given pressure, i.e. when ice becomes meltwater. Inside a glacier, because of the pressure, the melting point will be lower than 0°C. So it is significant as the threshold between two sets of processes – ice or running water.

12 The threats include farming, forestry, mining and quarrying, HEP and tourism. Tourism because tourists themselves damage the environment (trampling) and the development of tourism requires the creation of an infrastructure of hotels, cabins, ski slopes and ski lifts, access roads, etc. All these developments threaten to cause large-scale environmental damage.

13 The two extremes are 'do nothing' (business as usual) and 'total protection'. In between, the options are sustainable exploitation of some resources and comprehensive conservation. The former would allow some developments, such as carefully regulated ecotourism or organic eco-farming.

14 A high-energy coast is one mainly undergoing erosion, whereas a low-energy coast is one characterised by deposition.

15 Metamorphic rocks are rocks (mainly sedimentary) that have been changed in texture, composition or structure by either intense pressure and/or heat. Igneous rocks are formed by the solidification of magma either within the Earth's crust or at the surface. Sedimentary rocks are usually deposited in layers or strata, the material being largely derived from the breakdown of other rocks (i.e. igneous or metamorphic).

16 Sedimentary rocks because of their bedding planes and often loosely consolidated nature are most prone to erosion. Igneous rocks are hard and resistant to erosion. Metamorphic rocks are somewhere in between, but where folded and fractured can be very susceptible to erosion.

17 It supplies the sand that accumulates to form the dunes.

18 They are moving in a circular rotation.

19 From the erosion of the coast and from silt and sand brought down to the coast by rivers. Some may also come directly from weathering and mass movement. Material can also be rolled towards the coast from offshore deposits.

20 That is when wave energy is at its strongest and waves strike the coast with great force. Wave activity at this time will be predominantly destructive rather than constructive.

21 The cave is subjected to erosion, particularly hydraulic action. The cave is gradually enlarged until it reaches a cave being excavated on the other side of the headland. When they join, an arch is formed. This is gradually weakened until the point is reached when the cave roof collapses and the outer limb of the arch becomes a stack.

22 Yes, because it flows consistently in a given direction. But unlike ocean currents, it is not driven by differences in water salinity and temperature. It provides a means of transporting sediment along the coast.

23 Strong chemical weathering

24 In cold, temperate or periglacial regions, where there is little precipitation but temperatures fluctuate around freezing point. Much freeze–thaw activity causes much mechanical weathering.

25 a) Rockfalls are a rapid form of mass movement (rocks often falling vertically) whereas rotational slides are the outcome of mass movement along curved failure surfaces.
b) Screes are masses of weathered, angular rock fragments collecting below a free face to create a talus. Cliff terraces are the result of rotational sliding.

26 Barrier islands are low, sandy islands, usually in a chain-like formation parallel to the mainland. Formed by post-glacial sea-level rises, formerly continuous bars have been breached to create a string of islands.

27 Because they starve deltas of the sediment needed to maintain them. The equilibrium between deposition and erosion is changed in favour of the latter.

28 Coincidence of very low pressure and strong winds; powerful waves and swell; spring high tides; coastal morphology funnelling in the same direction as the wind.

29 Yes, but the scale and speed of coastal change are also considerations that need to be taken into account.

30 Sediment cells are natural divisions along the coastline that function as open systems. There are 22 around the coast of England and Wales. Any change within a cell is likely to have consequences within that cell. Because of this 'wholeness', a more joined-up form of coastal management is possible.

Chapter 3

1 The transport developments include railways, jet aircraft, container ships and the motor vehicle. They have greatly reduced the time taken to move goods and people between two locations. This has heightened connectivity and led to time–space compression.

2 Three IGOs have been particularly influential through the promotion of free trade and FDI: the IMF, World Bank and WTO.

Various agencies of the UN have encouraged the 'globalisation' of education and healthcare.

3 By:
- exploiting new reserves of resources
- setting up branch factories and offices around the world
- encouraging global trade in goods and services
- creating jobs in all four sectors
- disseminating a global culture.

4 A country can remain switched of for a number of different reasons:
- a political wish not to participate in the global economy, e.g. North Korea
- a remote or inaccessible location, e.g. land-locked Zambia
- little to offer by way of resources (human and physical), e.g. countries of the Sahel.

5 The global shift in manufacturing has meant a loss of jobs in deindustrialising countries (the losers) and a gain in emerging countries (the winners). However, when it comes to environmental pollution, for example, de-industrialising countries are winners and emerging countries losers.

6 It has made a substantial contribution to the demographic growth of megacities. More so than immigration. This internal migration is the outcome of the perceived attractions of the megacity and the perceived limitations of rural life. Modern communications have made rural people much more aware of the megacity and its opportunities. The main benefit of this migration for the megacity is meeting some of its need for labour – essentially cheap and unskilled.

7 Because of their global reputation and the opportunities they provide in terms of employment and improved quality of life. This encourages low-waged international migration.

For many people there will be a sense of kudos associated with living and working in a global hub. It is this that has generated much elite migration.

8 A situation in which the 'health' of a country depends the 'health' of other countries. This applies particularly in an economic sense. So there is an in-built guarantee of help from others if things go wrong. Interdependence also applies in a military context, particularly within an alliance such as NATO.

9 After an initial enthusiasm for globalisation, awareness of its benefits is now being balanced by a growing awareness of its costs. These include:
- adverse environmental impacts related to resource exploitation, industrial development and urbanisation
- the rich nations becoming richer at the expense of the poor ones
- the spread of neo-colonialism
- the erosion of indigenous cultures
- the loss of national autonomy.

10 Globalisation is encouraging not just economic development, but also the wider development process. So it is important to include measures of this wider process, which has a significant social dimension. The HDI is now a widely used in making international comparisons.

11 For a number of different reasons:
- the possession of sought-after resources (natural and human)
- a geo-strategic location
- open invitation to TNCs
- government encouragement to become involved in globalisation.

12 Globalisation has encouraged high volumes of international migration, i.e. attracting migrants with different ethnicities and cultural

backgrounds from various parts of the world. The immigration cores thus become ethnic and cultural melting pots. The diversity can often lead to tensions and conflicts.

13 Governments can do so by controlling: migration, particularly immigration; access to the internet; foreign direct investment; TNC activities; international trade.

14 Local sourcing reduces the amount of transport needed to deliver commodities and products to the consumer. This saves on the burning of fossil fuels and encourages grassroots development.

15 Recycling can be viewed as the first step towards the more ambitious goal of a circular economy. It reduces the rate at which new natural resources are used. The recycling process does itself require the use of energy and water, however.

Chapter 4

1 Regeneration is the long-term upgrading of rundown urban and rural areas. It is a much more common process in towns and cities.

The need for urban regeneration in the UK is related to deindustrialisation and the shift to a post-industrial economy.

2 The primary and secondary sectors have declined. The economy is now dominated by the tertiary sector, but its dominance is being challenged by the quaternary sector and the newly emerging quinary or 'knowledge' sector.

3 Social inequalities relate to discrimination based on age, gender, health, education, etc. The inequalities are people-based. Service inequalities relate to situations where services are unequally available or accessible (compare urban and rural places). The inequalities are place-based.

4 Income (personal wealth) – money to spend on good housing, education, healthcare and non-essential goods and services. But it does not necessarily bring happiness or satisfaction.

5 Upwards trends in numbers employed, numbers in full-time employment and number of skilled jobs.

6 Only you know the answer to this question, but answer it as accurately as possible.

7 Sink estates; abandoned industrial estates; disused docklands; former mining settlements

8 People and groups may be marginalised and excluded because of their language, religion or customs, and especially by wealth.

9 Age; gender; ethnicity; length of residence; quality of life

10 By relaxing planning laws; giving tax concessions; offering incentives to provide particular services, such as affordable housing or leisure space;

11 Designating areas for special development (e.g. science or retail parks); sites made available at reduced rental and business rates; promoting the advantages the area has to offer — good transport links, well qualified/skilled labour force, pleasant environment, etc.

12 Only you know the answer to this question, but answer it as accurately as possible.

13 • Economic – more jobs, higher incomes, more inward investment, less deprivation
   • Social – improvements in life expectancy and health; reduced inequalities
   • Environmental – improvements in environmental quality (less pollution, fewer abandoned buildings and more amenity space)

14 • Providers – developers along with landholders
   • Users or beneficiaries – local businesses, local communities
   • Governance – national government, planners, local councils
   • Influencers – planners, local communities, along with action groups and political parties

15 Again, this will depend on the investigations that you have carried out, but the stakeholders will be any individual, group or organisation with a particular interest in the actions and outcomes of a regeneration, project.

16 Ethnicity is the mix of people drawn from different cultural backgrounds. Ethnic groups may be defined on the basis of language, religion and forms of dress. Some issues with ethnicity can be helped by a willingness on the part of ethnic minorities to become assimilated. Isolation in tight communities is inclined to antagonise some white British people. Another issue is the unfounded fears that these ethnic minorities might be 'stealing' houses, jobs and services.

17 England is much more densely populated than Scotland, Wales or Northern Ireland. Particularly high densities can be found between Manchester and Merseyside in the Northwest and London and the Southeast. There is a noticeable concentration of the Scottish population in the Central Lowlands. Wales shows an empty heartland, with much of the population concentrated in South Wales. In Northern Ireland, the demographic focus is around Belfast.

18 Natural change (increase or decrease) and migrational change (international migration balance). Despite concerns about immigration, natural increase accounts for the greater part of the UK's population growth.

19 Immigrants tend to be young adults, so they help swell the population in the 20–40 age range. They may also come with their children and so help broaden the base of the pyramid. They might subsequently invite their parents to join them, in which case they might help thicken the pyramid towards the top.

20 Their significance is a practical one. They are important parameters in the planning of housing programmes. They impact on the number and size of dwellings needed. It has to be said that marital status is of much less consequence today. There are many more partnerships (as opposed to married couples) than there used to be, and many have children. Also the increased incidence of single-parent households has to be taken into consideration.

21 Only you know the answer to this but answer it as accurately as possible.

22 Proximity to central services and possibly place of work. Housing might be cheaper in older, less well-maintained areas.

23 Young, single people or recent immigrants.

24 Young adults leave either for higher education (not to return) or for an urban-based job. This has a detrimental effect on services (fewer children needing schooling) and businesses (more difficult to recruit labour). Services and businesses close. The overall decline in prosperity and the quality of life persuades older people to move away.

25 This question is again directed at your two place studies and any relevant evidence that supports the existence of different views on the places. The reasons are largely to do with personal perceptions and who you are. Personal factors such as age, gender, educational qualifications, type of employment and level of remuneration are among the most influential.

26 The movement of people and employment away from major cities to smaller settlements and rural locations just beyond the city or to more distant, smaller cities, towns and even remote rural areas.

27 The main source became the EU, particularly the eastern countries of Poland, Slovakia and Romania. Being from the EU, they are allowed visa-free entry to the UK.

28 • Incidence of mixed marriages or partnerships.
   • Number of children of mixed ethnicity.
   • Residential dispersal from original ethnic concentrations.
   • Mixed ethnicity in particular types of employment.
   • Degree of involvement in the wider community.

29 Only you know the answer to this but answer it as accurately as possible.

30 • Assimilation – number of MPs from ethnic minorities.
   • Social progress – percentage who have a higher education qualification.
   • Housing provision – housing association activity.

31 • Providers – those in the driving seat largely through ownership.
   • Users – consumers who may be affected, either gaining or losing out.
   • Governance – those in control, not necessarily elected.
   • Influencers – those wishing to make an intervention or take some form of action.

32 Only you know the answer to this but answer it as accurately as possible.

Chapter 5

1 Because around two-thirds of it is locked away in ice sheets and glaciers.

2 *Direct:* it affects the rate of runoff and the amount of water becoming throughflow and percolating to become groundwater flow. In other words, it directly affects the speed with which water is being circulated. It also affects the capacities of aquifers.

Indirect: geology affects soil characteristics, which in turn affect vegetation and ultimately the speed with which water is moved around the drainage basin cycle.

3 • Evaporation – increased because of bare ground exposed to direct sunlight – higher temperatures, more evaporation.
   • Runoff – increased because surface runoff is not impeded by vegetation.
   • Evapotranspiration – reduced because there are no trees to facilitate this process.
   • Interception – reduced because there are no trees to facilitate this process.
   • Infiltration – reduced because of faster runoff.
   • Groundwater – reduced because more water is being moved by increased runoff.

4 Because more of the ground surface is impervious – concrete and tarmac. Water entering drains is moved more quickly to streams and rivers. Both result in a faster delivery of rainwater to the drainage network, at such a rate as to exceed the drainage capacities. Once these are exceeded, flooding ensues.

5 El Niño events involve the reversal of the normal directions of ocean currents. They involve the warming of ocean waters along the west coast of South America. La Niña is characterised by unusually cool ocean temperatures in the equatorial region of the Pacific. Both have a considerable impact on weather.

6 Because of the rather arid climate and the high standard of living, the per capita water demand is high. Particular demands include irrigation, personal showers, swimming pools, etc.

7 Wetlands were thought to be expendable and unhealthy. Better to reclaim them and put the land to some economic use. It is now realised that they perform a useful role in the contexts of flood control and shoreline protection. They also act as useful stores and as filters trapping and recycling nutrients, as well as having high biological productivity (fisheries, birdlife). So what has changed is our understanding of the natural environment.

8 More energy in the atmosphere means more storms and heavier precipitation. That means more water in circulation and possibly being circulated at a faster rate.

9 The amount of global warming and its impact on the climate, particularly the amount and the distribution of precipitation. Uncertainty about future levels of water demand – depends on population growth rates and rising living standards. Will the supplies be sufficient? Will the mismatch between water availability and water demand increase or decrease?

10 More frequent droughts and extension of the world's arid areas. Population growth and development in those parts of the world where there is little or no water surplus. Particular factors include population growth, rising living standards, industrialisation and the extension of commercial agriculture.

11 Water is really like any other commercial commodity – its price is governed by demand and supply. Where demand exceeds supply, the price will be higher than where the situation is reversed. The amount of precipitation varies spatially, and so too does the distribution of water demand. Unfortunately the two distributions do not match. It is this mismatch that causes the price of water to vary spatially.

12 Most likely, the first outbreak will be in the Middle East where there is much unrest. Water is in short supply because of the arid climate. Human survival there means high per capita water consumption. Also there are shared rivers that are important suppliers of water. The Tigris–Euphrates and Jordan are likely flashpoints, as well as international water pipelines.

13 Environmental sustainability is about ensuring the future availability of water that is unpolluted and is deemed to be safe water. Economic sustainability is ensuring a secure water supply to all users at an affordable price.

Chapter 6

1 • Atmosphere – carbon dioxide, methane
  • Hydrosphere – dissolved carbon dioxide
  • Lithosphere – limestone, fossil fuels
  • Biosphere – living and dead organisms

2 Sequestering is the long-term storage of carbon, whereas photosynthesis is the process by which plants capture carbon and then store it in their issues.

3 Carbon cycle pumps move the carbon vertically, whereas thermohaline currents move it horizontally.

4 Tropical rainforest

5 Biological carbon is carbon stored in the form of living and dead organic matter.

6 Open fires for cooking and domestic hot water. Heat to convert water into steam used to generate electricity. A propulsive power for transport.

7 Heating and/or cooling homes. More domestic appliances. Increased car ownership.

8 Nuclear = recyclable; hydro = renewable; natural gas = fossil fuel; oil = fossil fuel; coal = fossil fuel; biofuels = recyclable.

9 All are primary.

10 For example, Royal Dutch Shell, Chevron and Lukoil

11 Favourable – if relying heavily on imported fossil fuels; if likely to be low cost. Unfavourable – if there is strong public opposition; if exploitation costs are high.

12 Because the construction of the necessary infrastructure will consume energy – electricity; transport of components to build the farms or power stations.

13 It is not always the case that rising income necessarily increases environmental awareness. Not all developed countries are concerned about the environment.

14 The precise relationship between levels of carbon emission and rising temperature is not known. There is uncertainty over the resilience of the Earth's systems. Human uncertainties include population growth, effectiveness of international agreements and the increasing reliance on renewable sources.

15 In the present context, adaptation is any action that reacts and adjusts to changing climate conditions, e.g. water conservation and management, development of more resilient types of farming. Mitigation is any action that either reduces or eliminates the long-term risk and hazards of climate change, e.g. international agreement on carbon emissions, promotion of renewable energy.

Chapter 7

1 Economic power, because this has a direct bearing on military and political power. Economic wealth means money to be spent on troops and military equipment. These, in turn, mean greater political influence.

2 'Hard power' refers to the way that countries get their own way by use of force. 'Soft power' is the power of persuasion – this is often enforced through diplomacy and culture.

3 The reasons include:
   • post-war bankruptcy, meaning there was no money to run, or defend, colonies
   • the focus on post-war reconstruction at home meant that colonies were viewed as less important
   • anti-colonial movements, for example in India, grew increasingly strong and demands for independence could not be ignored.

4 A unipolar world is one dominated by one superpower, e.g. the British Empire or the US-dominated world of today. A bipolar world is one in which two superpowers, with opposing ideologies, vie for power, e.g. the USA and the USSR during the Cold War. A multi-polar world is more complex: many superpowers and emerging powers compete for power in different regions.

5 There will be less of a cost of an ageing population, there will be a youthful, vibrant labour pool available and they will be able to adapt to new skills and technologies.

6 Because they move at different speeds along the economic development pathway. Those leading the pack are the superpowers. Those that slip back lose their status. Also changed by the creation of new alliances, such as the EU.

7 The dominance of TNCs in the global economy has been caused by a number of factors:
   • Their economies of scale mean they can outcompete smaller companies and, in many cases, take them over.
   • Their huge funds allow them to take advantage of globalisation by investing in new technology.
   • The move towards free-market capitalism and free trade has opened up new markets, allowing them to expand.

8 Global policing is needed to enforce international law. International peace and political stability are the prerequisites of a prosperous and expanding global economy.

9 Again, as in 1, the choice is most likely to be the economic pillar. Unfortunately the organisations involved in the three other pillars are proving to be weaker than the economic organisations such as the World Bank and WTO.

10 Because their current economic development depends heavily on cheap fossil fuels, particularly coal (the worst for carbon emissions). They might argue that many developed countries, in their time, burnt huge amounts of coal and were not pressured to turn to cleaner sources of energy.

11 The middle class traditionally combines education, modest wealth and ambition. But, more importantly, it has proved to be a source of enterprise and innovation, both particularly critical to a country's progress along the development pathway.

12 The ties could be through:
   • neo-colonialism – superpowers pulling the economic and political strings of developing countries, despite not ruling them directly, as during the colonial/imperial era
   • unfair terms of trade – cheap commodity exports for the developing world (coffee, cocoa, oil, copper) set against expensive manufactured imports from developed countries
   • the brain drain of skilled workers from developing countries to boost developed world economies
   • local wealthy elites, who control imports and exports in developing countries, benefiting from the neo-colonial relationship but having no interest in changing it.

13 Inevitably there is keen competition to become not just the top and most powerful economy in Asia, but also to become the regional superpower. Political power lies mainly in economic strength. Political tensions are intensified by ideological differences with, for example, China representing communism and India democracy.

14 There are many tensions in the Middle East related to oil and water resources, territory, and religious and ethnic differences. The issue for the superpowers is whether or not to make an intervention, particularly in the current conflicts. The sought-after prize is guaranteed access to the rich oil and gas resources. Emerging powers are simply concerned that oil supplies should not be disrupted.

15 The USA spends more per capita on defence than most countries. Certainly it is spending more than those countries it feels it is its duty (as the leader of the capitalist world) to protect. It is only natural that US citizens should ask the question: is superpower status really worth this expenditure? Why should others not be paying more?

16 • China challenging the USA as the leading global economy.
   • The shift of the global economy's centre of gravity to Asia.
   • The regional rivalry of China and India.
   • A new Cold War between the West and Russia.

Chapter 8

1 The HDI because it is based entirely on quantitative data. The HPI relies partly on some qualitative input.

2 Environmental quality has an impact on health, for example a polluted environment will have a detrimental impact. Health, in turn, finds expression in life expectancy. Longer life expectancy means more time to enjoy the benefits of a good quality environment.

3 There are many aspects, including the fact that it improves the quality of the labour pool, increases aspirations for a higher standard of living and better quality of life, gives better awareness of the prerequisites of healthy living, and increases the likelihood of better governance.

4 Number of doctors per 100,000 people; healthy life expectancy; percentage of population vaccinated for particular diseases.

5 • Globally – differences in diet, healthcare and quality of life.
   • Internationally – lifestyles, diet and economic prosperity.
   • Internally – ethnicity, socio-economic class, income and housing.

6 • World Bank – poverty reduction and support for development projects.
   • WTO – the liberalisation of trade.
   • IMF – encouragement of free market economies.
   • UNESCO – the promotion of education, protection of human rights and heritage.
   • OECD – the improvement of economic and social well-being.

7 Improving the quality of life today without damaging the quality of life of future generations.

8 ODA as a percentage of GNI. ODA needs to be put in a common context if countries are to be compared in order to determine whether they are pulling their weight. ODA for a large and prosperous country, such as the USA, may be impressive in absolute terms but minuscule when related to its wealth. While the UK's ODA is significantly smaller than that of the USA, proportionately it is significantly larger.

9 Because:
   • IS is intent upon further destabilising the Middle East and North Africa
   • it is a threat to global oil supplies
   • of its horrific abuse of human rights.

10 Torture is what is done, physical and mentally, to extract information. Rendition is sending someone to be tortured in a country which is lax in its enforcement of international laws that declare torture to be illegal.

11 Because human rights have an intangible quality. The best measures are negative ones, namely statistics relating to the abuse of human rights. The problem is that human rights are often subtly denied rather than overtly abused.

12 Gini coefficient

13 The USA, the EU, China and possibly Russia

14 It would be expensive in terms of human lives and damage to property. Military equipment is hugely costly to deploy and replace. Generally speaking, the longer the intervention, the longer the recovery period.

15 In theory, the demand for labour is greatest in the emerging countries, particularly China and India. But both these countries have huge labour resources of their own and do not rely on immigrant labour. Elsewhere, the demand for labour is high in the core regions of the world, such as the USA and the EU. One aspect that is new is the mobility of labour, thanks to modern transport and communications, as well as more open borders.

16 A political refugee is someone suffering some form of persecution or discrimination to the extent of genuinely fearing for their life. An environmental refugee is someone who has been forced to move because of some environmental event, such as a natural hazard.

17 Congestion; higher costs of living, especially housing; deterioration in environmental quality; peripheries drained to the point of disaster, etc.

18 Advances in modern transport have greatly changed time–distance values and the costs of long-distance travel. Modern communications now allow instantaneous contact almost everywhere in the world. The significance is to draw countries closer together, creating a feeling of belonging to a global village. It has also encouraged the growth of a global economy and national interdependence.

19 For: much easier movement of people, particularly economic migrants and refugees. Against: migrants are free to converge on favoured locations, causing all manner of

problems – excessive growth, aggravating social and ethnic tensions, etc. National unity can be threatened.

20 Because of the number of civilians and military personnel killed during the war and the huge amount of reconstruction work needing to be done. Main employers were public transport and the building industry.

21 The choice is yours! The negatives you have to weigh up are authoritarian control, corruption, economic bankruptcy and human rights abuse. Not an easy choice.

22 To maintain peace between internal factions; to protect human rights; to support elected governments; to sort out failed states.

23 • Ozone layer – difficult to monitor compliance.
   • Carbon emissions – compliance is a threat to the development of some countries.
   • Endangered species – difficult to police.
   • Biodiversity – poor awareness of ecosystem services.

24 Rivers that are shared by two or more countries. There are two possible forms of 'sharing': where the river forms an international boundary – e.g. the Uruguay River between Brazil, Uruguay and Argentina; and where the river's length is divided (the advantage being with the upstream partner) – e.g. the Ganges shared by India and Bangladesh, the Tigris by Turkey and Iraq.

25 Spillages and discharges from ships; trampling at landing sites; disturbance of penguin colonies; increase in the pressure to provide more land-based facilities, even hotels.

26 The growth of a materialistic consumer society; increasing multi-ethnicity; increasing exposure to a global culture.

27 Ethnic and cultural differences; political polarisation; acute core–periphery differences; widening gap between rich and poor.